A to Z
In Health &
Social Care

By

V. K. Leigh

authorHOUSE®

AuthorHouse™ UK
1663 Liberty Drive
Bloomington, IN 47403 USA
www.authorhouse.co.uk
Phone: 0800.197.4150

Published by AuthorHouse 03/15/2016

ISBN: 978-1-5246-2950-2 (sc)
ISBN: 978-1-5246-2951-9 (hc)
ISBN: 978-1-5246-2949-6 (e)

Print information available on the last page.

Any people depicted in stock imagery provided by Thinkstock are models,
and such images are being used for illustrative purposes only.
Certain stock imagery © Thinkstock.

This book is printed on acid-free paper.

Because of the dynamic nature of the Internet, any web addresses or
links contained in this book may have changed since publication and
may no longer be valid. The views expressed in this work are solely those
of the author and do not necessarily reflect the views of the publisher,
and the publisher hereby disclaims any responsibility for them.

Scripture quotations marked KJV are from the Holy Bible, King James Version
(Authorized Version). First published in 1611. Quoted from the KJV Classic
Reference Bible, Copyright © 1983 by The Zondervan Corporation.

Contents

How to use this book?

This is an illustrated book in a dictionary format set in alphabet i order of A to Z in health and social care; accompanied with quotations from the "New King James Version" of the bible.

The words contained in this book are used in the environments of Sheltered Housing, Domiciliary Care, Community Care, Social Housing, Extra Care, Social Care and other related fields.

This is an "Illustrated A to Z health and social care book with words and meanings cross referenced".

The format, to which it has been arranged, made it easier to use; coupled with words of wisdom for meditation

The contents in this illustrated A to Z book of health and social care are simplified with meanings to give simple understanding to the words.

This book can be referred to at anytime and anyplace making it easier to eliminate doubts from the mind while at work or wherever; some of these words are accompanied by diagrams and in some cases, real photographs are shown.

This illustrated A to Z in health and social care book will aid you *"the reader"* to locate the true meanings to your unanswered words with fewer efforts

Some quotations from the books of Proverbs from the New King James version of the bible have been included to make the findings and research more knowledgeable and interesting for reading and understanding.

I have enjoyed compiling this illustrated A to Z book in health and social care; hope you will find it useful and advantageous; with my nursing experience and the training of pupils in health and social care education.

"Look unto the day of Justice with peace"

About This Book

This is an **A to Z dictionary** type of glossary-book that contains the everyday words and vocabulary used in the environments of Social Care, Sheltered Housing, Extra Care, Social Housing and related environments.

These are the words that people use every day and are taken for granted without thinking about what they mean or represent.

Most of the staffs, clients, patients and visitors use these words so frequently and indiscriminately that to some people; they have lost the true meanings and understandings.

This book is of educational value to training establishments and colleges that educate pupils in health and social care, social housing, extra care, sheltered housing, care sectors; it is a source of reference, when unsure about the terms of words that are being used or pronounced.

It is important to realise that this is not a text book but only a reference source – *that is a* **"Dictionary".**

This **A to Z** dictionary format is a book that can be used on daily basis; useful during studies and in the place of work.

This book is set up in alphabetical order; each section starts with "words of wisdom" (*quotations from the book of proverbs*): the facts and the usefulness of these information come very handy, knowledgeable and educational.

This is a "***must have***" book, on "book-shelf" at home and at place of work for "*easy*" referencing.

"The way of progress is neither swift nor easy"

"A wise *man* will hear and increase learning; and a man of understanding will attain wise counsel" (Proverbs 1:5)

A&E – *(Accident & Emergency)* The accident and emergency (A&E) department at a Hospital provides a 24-hour emergency service, 365 days a year. The service has a dedicated A&E team, providing separate facilities for the treatment of children up to 16 years of age

AIDS – *(Acquired Immune Deficiency)* – decease of the blood caused by the Human Immune Virus (HIV); can be contacted by injecting with infected needles, sexual contacts and infected body fluids.

Abdomen – That part of the body containing the intestines

Abdominal Aortic Aneurysms - "AAA" or "Triple A," the most common form of aortic aneurysm, involve that segment of the aorta within the abdominal cavity; enlargement of the abdominal aorta such that the diameter is greater than 3 cm or more than 50% larger than normal.

They usually cause no symptoms except when ruptured

Abortion - The termination of pregnancy by the removal from the uterus of a foetus prior to viability; can occur spontaneously, in which case it is usually called a miscarriage, or it can be intentionally induced.

The term *Abortion* is the induced termination of a human pregnancy.

Ability – The confidence of being able to do something (work)

Abuse – To reproach, unjust, harsh, an evil practice that can cause harm or injury; action that intentionally harms or injures

1

another person directly or indirectly: there are several types of abuse: *physical abuse, sexual abuse, substance abuse, elder abuse, and psychological abuse;* to hurt or injure by maltreatment; ill-use.

Access – The right to information (personal files, medical records etc.)

Access to information – It (*Information Act)* provides the right of access to all information under the control of a government or institution be made available to the public, but with exceptions to the right of access which should be limited and specific; that decisions on the disclosure of any government information must be reviewed independently

Access to services - *Disability Discrimination Act (DDA)* the rights in the way a person can use services; that it is unlawful for service providers to treat anyone less favourably because of their disability, 'reasonable adjustments' must be made for the individual, in such that, changing the way they provide their services and render extra assistance

Accommodation – Space in a room to lodge in

Accuse – To bring a charge against someone.

Acid Erosion – A condition that affects the teeth, also known as **dental erosion**, is the irreversible loss of tooth structure due to chemical dissolution by acids not of bacterial origin; the damage caused by acids — often from food sources — softening the surface of the tooth's enamel, which is then more easily worn away.

Acknowledge – admitting receipt of.

Acne - skin condition that occurs when your hair follicles become plugged with oil and dead skin cells. Acne usually appears on your face, neck, chest, back and shoulders. Acne in age 14-year-old male/female during puberty; Acne occurs most commonly during adolescence

Active listening – Close listening, accompanied by an awareness of non-verbal communication of self and others.

Active participation – Ways of working and seeing everyone else and the client concerned as being active partners

Activity Based Interaction (ABI) – Activities designed to promote and develop communications and inter personal skills (In large groups, small group activities and on one-2-one) based on oral presentations and discussions

Acupuncture – Is a form of ancient Chinese medicine where fine needles are inserted into the human body at certain points; believing that energy or "life- force" flows through the body in channels that is called meridians. The life force is called qi (*pronounced "Chee"*)

Acute respiratory distress syndrome (*ARDS*) - Is a life-threatening lung condition that prevents enough oxygen from getting to the lungs and into the blood.

Acquired disorder - Is a medical condition which develops in contrast with a congenital disorder present at birth. A congenital disorder may be antecedent to an acquired disorder (such as Eisenmenger's syndrome)

Adolescence (Latin: *adolescere* meaning "to grow up") - Is a developmental transition between childhood and adulthood; the period from puberty to adulthood when human development has been fully attained

Adrenal glands – The adrenal glands are found directly above the kidneys in the human body, (suprarenal glands) are endocrine glands that rest at the top of the kidneys in humans, the right adrenal gland is triangular shaped, while the left adrenal gland is semi lunar shaped.

The Adrenal Glands are mainly responsible for releasing hormones in response to stress; adrenaline is one of several hormones produced by these glands. Sometimes called "fight or flight" is often used to characterize the body's reaction to very stressful situations.

Adrenaline -A hormone produced by the adrenal glands in high stress and exciting situations. This hormone is part of the human body's acute stress response system, sometimes called the "fight or flight" response. It works by stimulating the heart rate, contracting blood vessels, and dilating air passages; thereby increasing the blood flow to the muscles and oxygen to the lungs; used as a drug to treat cardiac arrest and other cardiac dysrhythmias resulting in diminished or absent cardiac output

Adrenocorticotrophic hormone (*ACTH*) - Is a hormone that stimulates the production of Cortisol, a steroid hormone useful for regulating glucose, protein, and lipid metabolism, thereby suppressing the immune system's response, and helping to maintain blood pressure. Located below the brain in the centre of the head, ACTH is produced by the pituitary gland, the pituitary gland being part of the endocrine, and a network of glands that work together to produce hormones that act on organs, tissues, and other glands to regulate systems throughout the body.

Addiction – The compulsive need for and use of a habit-forming substance (*as heroin or alcohol*); persistent and compulsive use of substances known by the user to be harmful

Addison's disease - Is a disorder that occurs when the ***adrenal glands*** produce fewer hormones.

Adulthood - Is commonly thought of as beginning at age 21; the Period in the human life span when full physical and intellectual maturity is attained; someone that is of reproductive age a person who has attained the age of majority and is therefore regarded as independent, self-sufficient, and responsible

Adults at risk – Any person in need of community care services that is aged eighteen years and over; someone who may have been subject to abuse or exposed to risks

Advice - Recommendation offered as a guide to action; recommending to someone about what should be done

Advisory Conciliation and Arbitration Service – (*ACAS*) is a Crown non-departmental public body of the Government of the United Kingdom. Its purpose is to improve organisations and working life through the promotion and facilitation of strong industrial relations practice; it gives advice on workplace issues including flexible work, sick pay, maternity and help with employment relations by supplying up-to-date information, independent advice and high quality training; working with employers and employees to solve problems and improve performances.

Advocacy – The process whereby a social worker or volunteer can act or speak on behalf of a client; the act of interceding on behalf of another person, taking their rights into consideration and to provide active support

Advocate – Someone who speaks in a supportive and official capacity on behalf of someone else.

Aerobic respiration (*ATP*) - Is the chemical energy "currency" of the cell that powers the cell's metabolic activities; this process is called *aerobic respiration* and is the reason animals breathe oxygen; the process of cells using oxygen to break down molecules and create energy.

Affection – Friendship to love; the affection of a parent for an only child. A moderate feeling or emotion

Affirm – To state positively; to maintain as the truth

Affirmative action – The kind of action that is taken in favour of those people discriminated against

Age - The length of time that one has existed; the duration of life; the period of human life, measured by years from birth

Ageism *(Disablism)* – Someone being less properly looked after because they are old or vulnerable

Aggression – Hostile behaviour, quarrelsome; a forceful, hostile or attack that occurs either in retaliation or without provocation.

It is an intention to cause harm or intent to increase relative social dominance.

Agony – To suffer excruciating pain

Agoraphobia – A fear of being outside; an anxiety disorder in which a person has attacks of intense fear and anxiety

Agreement – A decision accepted by everyone

Aid – Help; To Assist

Ailment – An Illness

Air Lift – To carry, lifting by Air

Air passages - Duct that provides ventilation (as in mines) air duct, airway duct - an enclosed conduit for a fluid register - an air passage

Ajar - Slightly open; partially open

Alarm – Device which makes sound to warn of danger

Albino – A skin due to lack of colouring pigment

Alcoholic –Someone that loves and drinks a lot of alcohol

Alcoholics Anonymous (AA) – A group of men and women who assist one another to beat the habit of drinking

Alimentary canal, (digestive *tract*) - Is the mucous membrane-lined tube of the digestive system through which food enters the body and solid wastes are expelled. It extends from the mouth to the anus and includes the pharynx, oesophagus, stomach, and intestines.

Alert – Raising the awareness of possible danger as a warning

Allergy – Abnormal sensitivity to the body

Allocate – To set aside for a purpose

Allows Access – Grant access

Alopecia - The loss of hair from the head or body; it can mean baldness, a term generally reserved for pattern alopecia or androgenic alopecia

> ➢ **Alopecia areata monolocularis** describes baldness in only one spot. It may occur anywhere on the head.
> ➢ **Alopecia areata multilocularis** refers to multiple areas of hair loss.
> ➢ Hair lost more diffusely over the whole scalp, is called **diffuse alopecia areata**

Alveolus - Cavities or sockets of the jaw, in which the roots of the teeth are embedded.

Alzheimer's disease – A disorder which has a degenerative effect on human such as mood swings, difficulty in remembering things, inability to remember someone, confusion and disorientation; the most common form of Dementia is Alzheimer's Disease. The symptoms of Alzheimer's disease build up gradually over a period of years.

Ambulance – Vehicle used to carry the sick to hospital

Ambulance Service – Deals with emergencies and taking clients to and from hospitals also as a form of transportation to hospital patients.

Ambulation - *am·bu·la·tion* (am'byū-lā'shŭn)

The activity of walking about: a nursing outcome from the **N**ursing **O**utcomes **C**lassification (NOC) defined as ability to walk from place to place independently with or without assistive device

Amendment – To alter; to correct

Amlodipine - Amlodipine (Norvasc) is a drug used to treat high blood pressure (hypertension) and angina. Amlodipine is in a group of drugs called calcium channel blockers. Amlodipine relaxes (widens) blood vessels and improves blood flow.

Ammonia – Sharp smelling gas, it is soluble in water

Amnesia – The loss of a person's memory – temporary with no specific time of recovery

Amphetamines – Drugs used as stimulants to reduce weights; has side effects on long and short term usages (see medical journal)

Amplified – Made bigger; *(in this particular context)* a boosted and louder sound

Amputate – To cut off a part of the body in sickness as in disease or an accident

Amylopectin - a soluble polysaccharide and highly branched polymer of glucose found in plants It is one of the two components of starch, the other being amylose.

Anaemia – Deficiency of haemoglobin in the blood (shortage of iron in the diet, etc.)

Anaerobic respiration –Respiration that happens in two ways: Aerobic respiration is oxygen-based cellular respiration. Anaerobic respiration is cellular respiration that occurs without oxygen.

Androgen -*called "male hormones,"* - It is a group of hormones that primarily influence the growth and development of the male reproductive system; a group of hormones that play a role in male traits and reproductive activity, present in both males and females, the principle androgens are testosterone and androstenedione.

Andropause (*male menopause*) - Men with prostate cancer or testicular cancer can also have hot flashes, especially those who are undergoing hormone therapy with anti-androgens, also known as Androgen Antagonists, which reduce testosterone to castrate levels. There are also other ailments and even dietary changes which can cause it. Men who are castrated can also get hot flashes.

Aneurysm - a sac formed by localized dilatation of the wall of an artery, a vein, or the heart.

Angiogram - Is a type of an X-ray test that uses special dye and camera known as fluoroscopy to take pictures of the blood flow in an artery (aorta) or vein (vena cava). Angiogram can be used to look at the arteries or veins in the head, arms, legs, chest, back, or belly; using the Angiogram, a person can look at the arteries near the heart (coronary angiogram),lungs (pulmonary angiogram), brain (cerebral angiogram), head and neck (carotid angiogram), legs or arms (peripheral), and the aorta

Anhedonia - The inability to gain pleasure from normally pleasurable experiences; main symptoms of major depressive disorder (MDD); a characteristic of mental disorders including mood disorders, borderline personality disorder, **schizoid personality disorder** and schizophrenia; depressive illnesses are disorders of the brain.

Anhidrosis: The inability to sweat normally; the absence or the deficiency of one's ability to sweat.

Ankle Sprain – Injury to one or more ligaments in the ankle; it is usually found on the outside of the ankle.

Ant –A small insect, very industrious; social insects related to bees. It is an insect attracted by sweet smells especially of cakes

Anti-discrimination policies - This is a list of anti-discrimination acts (often called discrimination acts), which are laws designed to prevent discrimination.

Anti-rejection drugs - Daily medications taken by organ transplant patients to prevent organ rejection

Anorexia Nervosa – Illness that makes a person not to eat food; can cause heart failure if not detected on time (*see Bulimia*)

Annual – Once a year

Antibiotics - Types of medications that destroy or slow down the growth of bacteria

Antibiotic resistance - A form of drug resistance whereby some microorganism, usually a bacterial species, are able to survive after exposure to one or more antibiotics; pathogens resistant to multiple antibiotics are considered **multidrug resistant** (MDR) or superbugs. Microbes, rather than people, develop resistance to antibiotics.

Antidote – **solution** - (*from the Greek antididonai, "given against"*) - To counter against; a remedy used to neutralize the effects of poison.

Appraisal – An expert or official evaluation of someone's work and its progress monitored every six months to twelve months

Antimicrobial – A substance that kill the growth of microorganism, bacteria and fungi

Anus – The opening at the posterior of the alimentary canal through which faeces is discharged from the body

Aorta - The aorta being the largest artery in the body begins at the top of the left ventricle, the heart's muscular pumping chamber which pumps blood from the left ventricle into the aorta through the aortic valve.

The aorta is divided into four sections: The ascending aorta (rises up from the heart): The aortic arch (curves over the heart): The descending thoracic aorta (down through the chest) and the abdominal aorta which (begins at the diaphragm)

Aortic aneurysm - is a general term for an enlargement (dilation) of the aorta; Aortic aneurysms are classified by their location on the aorta. An aortic root aneurysm or aneurysm of the sinus of Valsalva, *see Valsalva Manoeuvre*

Aortography –The placement of a catheter in the aorta and injection of contrast material while taking x-rays of the aorta.

Appropriate Attitude - Discussing and understanding of directions and definitions without emotions which will also require the right attitude or the right mind

Arrhythmia - (**Dysrhythmia**) - An irregular or abnormal heartbeat; arrhythmias have no symptoms; common symptoms are palpitations or a pause between heartbeats

Arteries - Are tough on the outside and smooth on the inside; blood is pumped out through one main artery called the dorsal aorta
n artery has three layers: an outer layer of tissue, a muscular middle, and an inner layer of epithelial cells.

Arteriole -Is a small diameter blood vessel in the microcirculation that extends and branches out from an artery and leads to capillaries.

Arthritis – A disease which affects the body's joints and is linked with Rheumatism

(Rheumatoid arthritis and *Osteoarthritis*)- *see osteoarthritis*

Arthrography - Is a procedure involving multiple x- rays of a joint using a fluoroscope (special piece of x-ray equipment) which shows an immediate x-ray image. A contrast medium

(Iodine solution) is injected into the joint area, this helps to highlight structures of the joint.

Assessment – Methods by which a client's need is determine; Valuation of one's needs to allocate suitable accommodation (medical, health, special educational assessment needs, etc.)

Agreed – To be in one accord

Archive - Storage; place where old notes and work materials are kept for future references

Asperger syndrome - Asperger syndrome is a form of autism. It is a lifelong disability. Autism is often described as a 'spectrum disorder' because the condition affects people in many different ways. People with this condition are often referred to as having 'the triad of impairments' and have difficulties in three main areas, such as social communication, social interaction and social imagination

Asphyxia (*suffocation*) – A condition in which oxygen is stopped from reaching the body tissues due to interruption or excessive carbon dioxide in the body that could result in unconsciousness and often death and is usually caused by interruption, obstruction or damage to parts of the respiratory system

Assembly point – An agreed location where everyone is to meet when there is an emergency evacuation

Assessor - A person who evaluates the merits of work prepared as part of a course of study

Association of Residential Care (ARC) – A group which provides support and promote the day care provision for people with learning disabilities

Asthma – A disease of the lungs caused by constriction of the muscles; chronic long-term lung disease that inflames and narrows the airways; a disease that affects the lungs, causing repeated episodes of wheezing, breathlessness, chest tightness, and night times or early morning coughing

Ataxia - The inability to coordinate voluntary muscular movements that is symptomatic of some nervous disorders.

Atheroma – (atherosclerosis) - Fatty degeneration of the inner coat of the arteries: the root cause of various cardiovascular diseases such as angina, heart attack, stroke and peripheral vascular disease.

Attachment relationship – Strong emotional bond between two or more people

Attitude - An expression of favour or disfavour towards a person, place, thing, or event

Audiogram - The audiogram is a graph which gives a detailed description of your hearing ability and which can be described as a picture of your sense of hearing.

Auditory – The sense of hearing

Autism – Is a spectrum condition; it is a lifelong developmental disability that affects how a person communicates with, and their relationship to other people. It is *a disorder* with no known cause or causes; It affects each person in different ways and its most common characteristics are – The person have difficulties with changes; There is difficulty with communication; difficulty in social relationship

Autism Spectrum Disorder (ASD) - One of a distinct group of complex neurodevelopment disorders characterized by social impairment, communication difficulties, and restrictive, repetitive, and stereotyped patterns of behaviour. Other ASDs include autistic disorder, childhood disintegrative disorder, and pervasive developmental disorder not otherwise specified (usually referred to as PDD-NOS).

Autoimmune disorder - Is a condition that occurs when the immune system accidentally attacks and destroys healthy body tissue. There are more than 80 different types of autoimmune disorders; an autoimmune disorder may result in: *The destruction of one or more types of body tissues; abnormal growth of an organ; Changes in organ function*

The tissues in the body commonly affected are: -

> ➤ *Blood vessels*
> ➤ *Connective tissues*
> ➤ *Endocrine glands such as the thyroid or pancreas*
> ➤ *Joints*
> ➤ *Muscles*
> ➤ *Red blood cells*
> ➤ *Skin*

Automatic Pill Dispenser (APD) - For home or institutional use Easy to use for all patients Alarms can be set for 1, 2, 3... Up to 28 times per day; the Ultimate Pill Dispenser for Alzheimer & Dementia patients It dispenses the Right Medication with the Right Dose at the Right Time. It has a Solid lockable lid so the patient is not tempted to access the medication before the correct time. The APD is programmed to dispense 1-2-3 or 4 times per day with Audio & Visual alarms Easy to Use - Easy to set up - 28 extra-large pill compartments will hold at least 12 aspirin sized pills with Spare trays & keys available for safety measures.

Autonomic Nervous System - That part of the vertebrate nervous system which regulates involuntary action, such as the intestines, heart, and glands, and that is divided into the sympathetic nervous system and the parasympathetic nervous system.

Autopsy – The operation carried out on a dead body to find out the cause of death

Autosomal - Refers to any of the chromosomes other than the sex-determining chromosomes (i.e., the X and Y) or the genes on these chromosomes

Autosomal dominant - A gene on one of the non-sex chromosomes that is always expressed, even if only one copy is present; Biotechnology Information; Autosomal dominance is a pattern of inheritance characteristic of some genetic diseases. "Autosomal" means that the gene in question is located on one of the numbered, or non-sex, chromosomes. "Dominant" means that a single copy of the disease-associated mutation is enough to cause the disease

Awareness - The ability to perceive, to feel, or to be conscious of objects, or sensory patterns.

*B*e not wise in thine own eyes: fear the LORD, and depart from evil". (Proverbs 3:7)

BPH - common problem for men aged 50years and over, is an enlarged prostate (BPH).

Benign prostatic hyperplasia (BPH), also called benign enlargement of the prostate (BEP), adenofibromyomatous hyperplasia and benign prostatic hypertrophy, is an increase in size of the prostate. A healthy human prostate is said to be slightly larger than a walnut.

BPH involves hyperplasia of prostatic stromal and epithelial cells, resulting in the formation of large, fairly discrete nodules in the periurethral region of the prostate.

When the prostate gland located below the bladder in men produces fluid components of semen.

Over half of men aged 50plus have enlargement of the prostate gland. This condition is sometimes called benign prostatic hyperplasia or benign prostatic hypertrophy (BPH)

It is not known exactly why this enlargement occurs.

BPH is not cancer and does not cause cancer. Not all men have symptoms of BPH

Baby Shower – Originally offered for the first baby of the mother-to-be but the tradition changed to being for all "new additions" to a family since they are all miracles and deserve to be showered with love and gifts. A ritual performed by well-wishers to a pregnant woman about to deliver a baby; a forbidden ceremony for the true church of God.

Back pain – A common condition; it can be very uncomfortable but not serious. It can affect anyone especially people aged between 35years and above

Bacteria – Bacteria are microscopic organisms whose single cells have neither a membrane-enclosed nucleus nor other membrane-enclosed organelles like mitochondria and chloroplasts. Some bacteria are good while some are bad for the human systems. There are four common types of bacteria namely, Vibrio bacteria; these are curved shaped and causes Cholera, Cocci bacteria: these are circle shaped and causes boil, Bacilli bacteria; these are shaped like a rod and causes typhoid fever, Spirilla bacteria; these have spiral shape and causes syphilis.

Bacterial Vaginosis *(BV) -Is the most common vaginal infection in women of childbearing age; a* smelling vagina is caused by Bacterial Vaginosis. An infection caused when bacteria change the normal balance of bacteria in the vagina; commonly affect women between the ages of 15 to 44.

BV is not considered a Sexually Transmitted Disease (*STD*) but having it increases the chances of getting an STD. BV cannot be contracted from toilet seats, bedding, or swimming pools.

Women with BV do not have symptoms and if they do, they will notice thin white or gray vaginal discharge, odor, pain, itching, burning or both in the vagina. Strong fish-like odour will be experienced, especially after sex. Burning will also be experienced when urinating; itching around the outside of the vagina, or both.

Bacterium – Micro-organism that carry diseases

Barbiturates – (*Tranquilizers*), commonly used to treat Insomnia (lack of sleep)

Barrier to communication - The barriers of effective communication can be a distraction from a fellow co- worker or pre-occupied mind; therefore, when an inflow of continuous messages come from different people and places, it is best to have a writing pad handy in order to write and pass messages on without the brains being overworked

Barrier contraceptive - Methods of contraception that literally create a barrier that stops the spermatozoa from reaching and fertilizing a woman's egg. Must be used as directed

Baseline – A form of behaviour; it is used to monitor and determine the kind of help and support that is given to a client (a form of assessment)

Baseline Observation – Measurements taken by a doctor, nurse or carer of blood pressure, temperature, pulse, etc. and recoded on a chart

Bed-Bug – Insect that infest houses, usually seen moving at night or in dark corners e.g. at corners of mattresses, corners of chairs or sofas; a creature that feeds by sucking the blood of its victims when sleeping

Behaviour - It is what we do and how we act; the range of actions made by organisms, systems, or artificial entities in conjunction with their environment, which includes the ways in which people are observed to conduct themselves

Beliefs - The assumptions we make about ourselves, about others in the world and about how we expect things to be

Benefits – This is a legal payment made to members of the community who need assistance financially

Benign – If a growth is benign, then it is non-cancerous

Benign Prostatic Hyperplasia (BPH) - also called benign enlargement of the prostate (BEP), adenofibromyomatous hyperplasia and benign prostatic hypertrophy, is an increase in size of the prostate.

BPH is a common problem for men over 50 is an enlarged prostate

BPH involves hyperplasia of prostatic stromal and epithelial cells, resulting in the formation of large, fairly discrete nodules in the periurethral region of the prostate

BEP - also called benign enlargement of the prostate; Benign prostatic hyperplasia (BPH), adenofibromyomatous hyperplasia and benign prostatic hypertrophy, is an increase in size of the prostate.

Bereavement – The death or loss of a loved one; the period of mourning and grief following the death of a beloved person or animal.

Bias – Tendency to a one-sided favour over someone else

Bigotry – A strong partiality to one's own group and being intolerant of those who are different

Bile – (**gall)** is a bitter alkaline fluid of a yellow, brown, or green colour, secreted, in man, by the liver. Bile, or gall, is composed of water, bile acids and their salts, bile pigments, cholesterol, fatty acids, and inorganic salts

Bilingualism - the ability to speak two languages; able to use two languages, especially with equal or nearly equal fluency.

Bind – To put together (a book)

Biopsy – Removal of tissues and its examination

Bipolar disorder - Bipolar disorder is a condition in which a person has periods of depression and periods of being extremely happy or being cross or irritable; it is characterized by a person's changing moods —from extreme highs (e.g., mania) to extreme lows (e.g., depression); also known as manic-depressive disorder; this is a mood related illness.

Bicuspid Aortic Valve - A bicuspid aortic valve is an aortic valve that only has two leaflets, instead of three.

Birbeck granules (**Birbeck bodies) -** rod shaped or "tennis-racket" cytoplasmic organelles with a central linear density and a striated appearance

Blister – Thin bubble on the skin surface caused by burns (hot water); a small pocket of fluid within the upper layers of the skin, typically caused by forceful rubbing (friction), burning, freezing, chemical exposure or infection. Blisters are filled with a clear fluid called serum or plasma; can be filled with blood (known as blood blisters) or with pus (if they become infected).

Bloat - known as **Gastric dilatation volvulus** or a **twisted stomach** - It is a medical condition in which the stomach becomes overstretched by excessive gas content; commonly referred to as **torsion** and **gastric torsion** when the stomach is twisted. *Bloat* is often used as a general term to cover gas distension of the stomach with or without twisting.

Blood – Red liquid flowing through the arteries and vein

Blood cells - Is a cell produced by haematopoiesis and normally found in blood. In mammals, these fall into three general categories: Erythrocytes (Red blood cells): Leukocytes (White blood cells): Thrombocytes (Platelets)

The diagram above illustrates the different types of blood cells. Leukocytes, also known as white blood cells, are a group of related cell types that involved in immune function. Leukocytes include neutrophils, eosinophils, basophils, lymphocytes and monocytes.

Blood Clot – When injured, soluble fibrinogen present in the blood plasma turns to insoluble fibrin (a thick blood that stops the bleeding)

Blood Pressure - the force of blood against the walls of arteries; a serious condition that can lead to coronary heart disease, heart failure, stroke, kidney failure, and other health problems

Blood system - Exists to inspire people to donate blood in order to produce a safe and ample blood supply; embracing continuous quality improvement

Body fluid – Fluids that circulates in the body or comes out from the body e.g. urine, blood, saliva, etc.

Body language – Communication expression through postures and gestures; non-verbal communication involving facial expressions that reveal someone's feelings

Body space – The physical space between two people that feels comfortable

Boil - A skin infection that starts in a hair follicle or oil gland. At first, the skin turns red and painful in the area of the infection, and a tender lump develops.

Bone - Bones are rigid organs that form part of the endoskeleton of human body. Skeleton and Human Body Bones, parts, organs, systems, organ system; **the** hard, rigid, connective tissue constituting most of the skeleton of vertebrates, composed mainly of calcium salts. There are more than 200 anatomically distinct structures making up the human skeleton.

Bone Marrow - The soft, flexible, vascular tissue found in the hollow interior cavities in the centre of many bones, the tissue comprising the centre of large bones; it is the place where new blood cells are produced; healthy bone marrow is essential for the body to function. The spongy tissue inside some bones, such as the hip and thigh bones contain immature cells, called stem cells.

Bowel – Intestines; the intestine or bowel is the segment of the alimentary canal extending from the *pyloric sphincter* (a ring of smooth muscle fibres around the opening of the stomach into the duodenum) of the stomach to the anus

Bowel Disorder – Intestine irregularity

Bradycardia - a heartbeat that is too slow - below 60 beats per minute *(see Tachycardia)*

Braille *(Moon)* – The methods by which sightless person can read using tactile dots or shapes; system which enables blind and partially sighted people to read and write through touch

Brain - The portion of the vertebrate central nervous system that is enclosed within the cranium, continuous with the spinal cord, and composed of cerebrum, midbrain and medulla oblongata (see medical dictionary for detailed information)

Brain tumour - An abnormal growth of cells within the brain. It is slow in growth; usually stays in one place and does not spread.

Breast – These are mammary glands, usually in pairs for human beings and more in other mammals; situated at the upper part of the chest; an important part of a woman's anatomy that produces milk to feed the new born baby.

Breathless – Difficulty in breathing

Brine – Mixture of water and salt; a salty solution that can be used as a preservative to inhibit bacterial growth.

Bribe – Gift to entice unaided action

Brief – As in statement, Short.

Brittle diabetes - Brittle diabetes, also known as labile diabetes, a term used to describe uncontrolled type 1 diabetes. Someone with brittle diabetes always experience swings in blood sugar levels. These cause either hypoglycaemia (low blood sugar) or hyperglycaemia (high blood sugar), which is more common and sometimes extreme.

Broken-down – Not in good condition as in health

Bronchitis – Inflammation of the wind pipes known as the bronchial tubes, characterized by coughing, difficulty in breathing, etc., and caused by infection or irritation of the respiratory tract

Brother – A name called by children of the same parent.

Brother-in-Law – The spouse of a sister

Brown Bread – Bread made with wheat

Bruise – Injury made by blunt object to the skin, causing it to change colour; Bruise (the medical term is contusion); injuries to soft tissue beneath unbroken skin

Older adults often bruise easily from minor injuries, especially injuries to the forearms, hands, legs, and feet. As a person ages, the skin becomes less flexible and thinner because there is less fat under the skin. Women bruise more easily than men, especially from minor injuries on the thighs, buttocks, and upper arms.

Brutal - To be Cruel

Brush – An object with bristle used for cleaning / washing

Bubble – Floating air in liquid – as in bubble bath

Bucket – Container or cylinder for holding water; a watertight, vertical cylinder or truncated cone, with an open top and a flat bottom, usually attached to a semi-circular carrying water or sand

Budget – List of how money is to be spent or spent

Bug – Electronic device used for listening

Bug (*Insect-like creature*) - A type of insect with mouth shaped like a straw which they use to suck plant juices from plants. The assassin bugs use their stylets (mouth shaped like straw) to suck blood from other insects e.g. milkweed bug and stink bug

Bulimia - Bulimia is an eating disorder characterized by overeating and purging *(see anorexia nervosa)*; this happens when purging is self-induced, vomiting, exercising oneself excessively and using laxatives.

Bullying - The use of force to abuse or intimidate others. This behaviour can be by habitual or physical power including verbal harassment or threat, physical assault directed repeatedly towards particular persons, perhaps on grounds of class, race, religion, gender, sexuality, appearance, behavior, or ability.

Bum – Buttocks; two rounded portions of the anatomy, located on the posterior of the pelvic region of humans or apes

Bump – A dull but heavy blow

Bun – A small cake

Buried penis - A buried penis occurs where the shaft of the penis is literally buried under excess skin and fat – this could be as a result of obesity and significant weight loss. It is not the size that matters but an issue when it is to be used; up to 1 out of every 200 men is born with what's medically known as 'micro-penis'; a normal sized penis that lacks an appropriate sheath of skin and is located beneath the integument of the abdomen, thigh, or scrotum; a congenital condition that can lead to the obstruction of the urinary stream, poor hygiene, soft tissue infection and inhibition of normal sexual function.

Burn – Injury caused by fire. Immense heat exposure

Bury – To put in a hole and cover from the eyes

Butter – Creamy flavoured fat used as spread on bread or for cooking

Button – Object used for fastening

By-pass - To avoid obstruction – as in by-pass operation of the hear

"Can a man take fire to his bosom; and his clothes not be burned?" (Proverbs 6:27)

CE *marking* – A trade mark conforming to the European standard, identifying health and environmental requirements

Carboplatin - is a chemotherapy drug used for treatment of many types of cancer; use for treating ovarian and non-small cell lung cancer, often use for other cancers such as Testicular, stomach, bladder cancers and other carcinomas.

Cardiomyopathies - diseases of the heart muscle itself; sometimes called an enlarged heart.

The heart is abnormally enlarged, thickened, or stiffened and as a result, the heart's ability to pump blood is weakened. This could result in heart failure.

Cardiovascular disease – (**CVD**) - a term covering diseases of the heart and blood vessels

Canadian Association on Gerontology *(CAG)* - is Canada's premier multidisciplinary association for those who research, work and have an interest in the field of aging.

Cancer - Cancer is the uncontrolled growth of abnormal cells in the body. It is named for the organ or type of cell in which it starts growing; a malignant neoplasm (tumour); is a broad group of various diseases, all involving unregulated cell growth; this is the most serious of all ailments. Over two thirds of the elderly are affected by this disease. Lung and breast cancer are the most common, with skin cancer making the occasional appearance. Cancer can be treated successfully nowadays, but the success rate is low in aged patients.

Cancer of the Womb – (***Uterine cancer***) - the most common invasive cancer of the female reproductive system

Cannabis - is also known as Ganja, grass, Hashish, Hemp, Indian hemp, marijuana, Pot, reefer, weed; most often consumed for its psychoactive effects which can include heightened mood, relaxation, and increase in appetite.

Its unwanted side-effects can sometimes include decrease in short-term memory, dry mouth, impaired motor skills, reddening of the eyes and feelings of paranoia or anxiety.

Capillaries - Are the smallest of a body's blood vessels and are parts of the microcirculation. Their endothelial linings are only one cell layer thick There are three main types of capillaries: *Sinusoidal capillaries: Fenestrated capillaries and Continuous* - They are continuous in the sense that the endothelial cells provide an uninterrupted lining

Cardiac catheterization - Involves passing a thin flexible tube (catheter) into the right or left side of the heart, usually from the groin or the arm; this test involves the collecting of blood samples from the heart; the Measurement of pressure and blood flow in the heart's chambers and in the large arteries around the heart; Measuring the oxygen in different parts of your heart; Examining the arteries of the heart and Perform a biopsy on the heart muscle

Cardiac tissue - (Cardiac muscle - heart muscle) - A type of involuntary muscle found in the walls of the heart

Cardio Pulmonary Arrest (CPA) - Cardiac refers to the heart - "pulmonary" refers to either lung function or the pulmonary artery that carries blood from the heart to the lungs and "Arrest" means to stop. - Cardiopulmonary arrest is a stoppage of the heart-lung function. It is known as a heart attack; **known** as Cardiac arrest or circulatory arrest, is a sudden stop in effective blood circulation due to failure of the heart to contract effectively. It is different from congestive heart failure, where circulation is substandard, but the heart is still pumping sufficient blood to sustain life. If cardiac arrest goes untreated for more than five minutes, injury is likely to happen to the Brain.

Capacity – The physical and mental ability to manage or do something

Care – Principles of good practice - Principles agreed for workers in health and social care; such as promoting values of equality and diversity, developing and maintaining relationships with clients but maintaining the boundaries, maintaining information confidentiality provided by clients, allow clients to be heard, freedom of speech, etc.

Carer – Someone who takes care of the vulnerable persons; a person of any age, providing support to someone unable to manage due to illness, disability, substance misuse or mental health.

Care Plans – An outline procedure which states the type of care to be provided by various professional carers to a client. This is reviewed and monitored periodically (NHS & Community Care Act 1990); detailed documents that states the needs and wishes of an individual, stating who will be responsible for the needs and wishes

Care Quality Commission – (**CQC**) –This is a body set up by the government of the United Kingdom; their role is to make sure that hospital, care homes, dental and general practices and other care services in England provide people with safe, effective and high-quality care; encouraging them to make improvements. Also responsible for the inspections, regulates health and social care services in England

Care value base - A set of rules of regulations and guidelines that every care practitioner has to adhere to in order to provide effective service to their clients or patients. There are seven principles involved, namely,

1	The Promotion of anti-discriminatory practice
2	Maintaining the confidentiality of information
3	The Promoting and supporting of the individual's right to dignity, independence, choice and safety
4	The Acknowledgement of people's personal beliefs and identities
5	Protecting the individuals from abuse
6	The effective provision of communication and relationships
7	The acknowledgement of Providing individualized care

Care Worker – The responsible person maintaining the care of a vulnerable person

Carrier - a person or thing that carries disease in or on their body without being infected

Case study - Is a detailed, in-depth study about a person, small group, or situation.

Cataract - A clouding of the lens inside the eye which leads to a decrease in vision. It can affect one or both eyes and develops slowly. The symptoms include faded colours, blurred vision and around the light halos, problems looking at bright lights, and difficulty seeing at night. It causes impaired vision; usually occur when there is upsurge of protein in the lens which makes the eye cloudy. It is the most common cause of vision loss in people over the age of 40; one of the principal causes of blindness in the world today.

Catheter – Designed for both male and female; thin, flexible tube; a tubular, flexible surgical instrument that is inserted into a cavity of the body to withdraw or introduce fluid; Condom catheters are most often used in elderly men with dementia; a flexible surgical instrument that is inserted into a cavity of the body to withdraw or introduce fluid.

Caucasians - *Caucasian race (Caucasoid or Europid)* has historically been used to describe the physical or biological type of some or all of the populations of Europe, North, the Horn of Africa, Western Asia, Central Asia, and South Asia

Cell - Is the basic structural, functional and biological unit of all known living organisms; Cells are the smallest unit of life that is classified as a living thing, often called the "building blocks of life". An adult human body is estimated to contain between 10 and 100 trillion cells

Cell division - The process by which a *parent cell* divides into two or more *daughter cells*; Cell division usually occurs as part of a larger "cell cycle"

Cell cycle - Is the series of events that take place in a cell leading to its division and duplication (replication). In cells without a nucleus, the cell cycle occurs via a process termed binary fission.

Central Nervous System – This is the centre that controls the body. It consists of the brain and the spinal cord.

Central Nervous System

The brain controls most activities in the body while the spinal cord deals with the nerves impulses from all areas of the body passing through the brain

Central Sleep Apnea – (CSA) - is a less common type of sleep apnea. This is a disorder which occurs if and when the area of the brain controlling the breathing doesn't send the correct signals to the breathing muscles; can affect anyone. Commonly affect people on medication with certain medicines. Snoring does not take place with **central sleep apnea**.

Cervical cap - (for female use) is a contraceptive device consisting of a small thimble-shaped cup that is placed over the uterine cervix to prevent the entrance of spermatozoa

Cervical spine – Age related changes that take place in the neck

Cervical Spondylosis (*Spinal Arthritis*)- A term used in the medical profession for the general wear and tear that takes place in the joints and bones of the spine as a person gets older; changes that comes with ageing affecting the neck (*cervical spine*)

Challenging behaviour – Problem behaviour which may take the form of shouting, biting, screaming, uncontrolled behaviour which could be a danger to society

Chancroid (*soft chancre* & *ulcus molle*) - Is a bacterial sexually transmitted infection characterized by painful sores on the genitalia; it is known to spread from one individual to another solely through sexual contact; Chancroid is caused by a type of bacteria called *Haemophilus ducreyi*.

The infection is found mainly in developing and third world countries

Infected men have only a single ulcer. Women often have four or more ulcers

Chart – A sheet exhibiting information in tabular form; graphic representation, e.g. curves, dependent variable, temperature, price, etc.

Charter standards - A kite mark, which recognises and rewards high quality levels of provision in club and business

Chest Infection - A chest infection is an infection that affects the lungs either in the bronchitis – larger airways- or in the smaller air sacs (pneumonia); there is a build-up of pus and fluid (mucus) and the airways become swollen making it difficult to breathe

Child protection register – Safeguarding, confidential records of names of children at risk

Chiropody – The practice of maintaining healthy feet; especially of older people and the vulnerable

Choice - The opportunity or privilege of choosing freely <freedom of *choice*>

Choose – To select, decide

Circulatory System -Is an organ system that allows blood and lymph circulation to transport nutrients (such as amino acids and electrolytes), oxygen, carbon dioxide, blood cells, etc. to and from cells in the body to nourish it and help to fight diseases; it stabilizes the body's temperature and maintain homeostasis

Circumcision - (*Latin circumcidere - "to cut around*) the surgical removal of the foreskin (prepuce) from the human penis; in a typical procedure, the foreskin is opened and then separated from the glans after inspection.

Classified – Secret

Claustrophobia – Fear of confined / closed places

Client – Is someone who is who is receiving the benefits, services, etc., of a social welfare agency, etc.

Clients' rights and choices - The right to make personal choices for their own healthcare, including the right to accept or refuse care

Clip – Fastener to hold together e.g. paper

Climb – To go up with difficulty

Clinic – A place of treatment

Clinical thermometer - (***medical thermometer***), is used for measuring human body temperature. The tip of the thermometer is inserted into the mouth under the tongue (*oral or sub-lingual temperature*), under the armpit ((*axillary temperature*), or into the rectum via the anus (*rectal temperature*).

Clinical waste – Medical waste; normally refers to waste products that cannot be considered general waste, produced from healthcare premises, such as hospitals, clinics, doctors' offices, veterinary hospitals and labs.

Chloasma (*Melasma*) - Melasma is a common skin disorder seen in men and especially women; commonly seen in pregnant women and often referred to as the mask of pregnancy. A patchy brown or dark brown skin discoloration that usually occurs on a woman's face

Chloasma frequently goes away after pregnancy. (***See Melasma)***

Closed questions – Questions that are worded in a way that invites a one-word answer, Yes or No.

Cobweb – A spiders spurned web

Codes of practice – The document which sets out the standards of practice; written guidelines issued by a professional association or an official organisation to its members in order to assist them to comply with its ethical standards

Codes of professional conduct for registered nurse, midwife or health visitor - As a registered nurse, midwife or health visitor, you are personally responsible for your practice. In caring for patients and client

  ➤ We should respect the patient or client as an individual
  ➤ We should obtain their consent as an individual before giving to them any treatment or care
  ➤ The protection and the confidentiality of client information
  ➤ To work and co-operate with others in the team record
  ➤ To maintain your professional knowledge and competence and ethics
  ➤ To be trustworthy by clients and patients
  ➤ To act to identify and minimise risk to patients and clients

These are the shared values of all the United Kingdom health care regulatory bodies

Code of professional conduct

This Code of professional conduct was published by the Nursing and Midwifery Council in April 2002 and came into effect on 1 June 2002.

Cocaine - Cocaine is a powerfully addictive stimulant drug made from the leaves of the coca plant native to South America. It produces short-term euphoria, energy and talkativeness in addition to potentially dangerous physical effects like raising heart rate and blood pressure.

Coerce – When someone is forced to perform a task or to do something against their desired will

Coffee – Drink made from roasted ground beans; roasted brown beans of a tropical tree used to make coffee. They are ground, or made into powder or granules that dissolve in hot water; the drink contains caffeine with a mild stimulating effect.

Congenital heart disease - a type of defect in structures of the heart or blood vessels that occurs before birth; this disease affects up to 9 out of every 1,000 children born in the UK and more so in other parts of the world.

Cognitive - A person with a cognitive disability has greater difficulty with one or more types of mental tasks than the average person; cognitive disabilities is a broad subject and not always well-defined – *see medical journal*

Cognitive Behavioural Therapy (*CBT*) – A type of psychological therapy, widely used for people with chronic back pain; based on the principle *"the way you think is partly the way you feel"*

Cognitive disorder – Conditions that affect a person's ability to use their memory or their ability to think e.g. amnesia and common among the older people, dementia, and Alzheimer's disease

Cohabit - To live together as if married, usually without legal or religious sanction; living together in an intimate relationship.

Collaborate – Treacherously working together

Colleagues – Persons associated with professionally or working together.

Collude – Secret agreement with someone to commit an undesirable offence

Colorectal cancer – (***colon cancer, rectal cancer, or bowel cancer***) is a cancer from uncontrolled cell growth in the colon, parts of the large intestine or in the appendix. Analysis shows that colon and rectal tumours are genetically the same cancer

Colostomy - surgical procedure in which a stoma is formed by drawing the healthy end of the large intestine or colon as it called through an incision in the anterior abdominal wall and suturing it into place; it drains stool (faeces) from the colon into the colostomy bag. Colostomy stool is often softer and more liquid than stool that is passed normally.

Thousands of colostomy surgeries are performed every year. The procedure of a colostomy requires that a portion of the large intestines be removed from the body. Then, a part of the colon is attached to the anterior abdominal wall, leaving a stoma (small opening) in its place. After this surgical process, a colostomy bag attached to the stoma is required to rid the body of faecal matter effectively.

Colostomy bags - Colostomy patients who do not irrigate have essentially the same choices available for those with an ileostomy.

Come to grips – Take a hold of one-self

Commission – Doing something deliberately, knowing what the consequences will be

Commission for Racial Equality - (CRE) - A non-departmental public body in the United Kingdom, aimed to address racial discrimination and promote racial equality; now the new Equality and Human Rights Commission

Community Care – Care offered by the community organisations e.g. Hospital and Agencies (Contract)

Community care plans - The care plan is a means of communicating and organizing the actions of a constantly changing nursing staff. As the patient's needs are attended to, the updated plan is passed on to the nursing staff at shift change and during nursing rounds.

Community Care – *Direct payment Act 1996* – The power to give disable people finances to buy their own social care services

Community Care - *Residential Accommodation Act 1998* - This allows the elderly the right to a savings of up to £16,000

Competence – Possession of required skill, knowledge, qualification, or capacity

Computer Axial Tomography - (CAT)) is a radiologic imaging modality that uses computer processing to generate an image (commonly known as *CT scan*) of the tissue density in a "slice" as thin as 1 to 10 mm in thickness through the patient's body; these images are spaced at intervals of 0.5 to 1 cm. Cross-sectional anatomy and can be reconstructed in several planes without exposing the patient to additional radiation.

Compliance - The tendency to yield to others, especially in a weak way

Communal – Shared in common by everyone in a group

Communication – Sharing with one another and yet with different opinions according to experiences and backgrounds

Condom - A flexible sheath, usually made of thin rubber or latex, designed to cover the penis during sexual intercourse for contraceptive purposes or as a means of preventing sexually transmitted diseases.

Confidence – Self-assurance; a state of being certain; reliability of a person

Confidentiality – The requirement to keep personal information private and only share it with those who need it; ensuring that information is only accessible to people who are authorised to know it

Congenital disorder - Is a medical condition that's present at birth; it is an abnormality of the structure of a body part, it may or may not be perceived as a problem condition

Congenital Insensitivity to Pain with Anhidrosis (CIPA) - Is a hereditary sensory and autonomic neuropathy type IV Condition; the inability to feel pain and temperature decreased or absent sweating (***Anhidrosis***). The signs and symptoms of CIPA appear early, usually at birth or during infancy, but with careful medical attention, affected individuals can live into adulthood. Some people with CIPA have weak muscle tone (***hypotonia***) when they are young, but muscle strength and tone become more normal as they get older.

Conjunctivitis (also called **pink eye** in America or **madras eye** in India) – is an inflammation of the outermost layer of the eye and the inner surface of the eyelids. Caused by an infection usually viral, but sometimes bacterial or an allergic reaction

Connective tissue (CT) - Is a biological tissue which supports, connects, or separates different types of tissues and organs of the body.

CT has three components: cells, fibres, and extracellular matrices, immersed in the body fluids

One of the four general classes of biological tissues—the others are epithelial, muscular, and nervous tissues; it is estimated that 1 out of 10 people have a Connective Tissue Disorder.

Consent – Allowing something to happen with full permission; there are different types of consent within health and social care environment – ***Informed consent*** *(the individual is fully aware of the outcome of the decisions),* ***Continued consent*** *(informed consent is effective during care or support),* ***Consent by proxy*** *(where decisions are made by someone else e.g. family member)* and ***Implied consent*** *(impression of consenting with no formal agreement e.g. a doctor)*

Constipation - Constipation is usually caused by a disorder of bowel function rather than a structural problem. It occurs among all ages, from new-borns to elderly persons. However, this condition is more common among the elderly. Women experience this problem more than men.

Constructive - advice that is useful and intended to help or improve something, often with an offer.

Contamination – The infection of an object or on a person

Contraceptive - any device that prevents or tends to prevent conception

Contrast Media Techniques - To allow the visualization of the outline of structures or organs, which otherwise would be difficult to demonstrate clearly, Contrast media are employed in radiography most commonly used for brain scan and X-ray of internal organs

Contribute - To give in common with others

Control of Substances Hazardous to Health (*COSHH*) 2002 - Employers to take the necessary precautions to ensure that hazardous substances are stored and used correctly; an independent organisation that regulates health and social care services.

In order from top left to bottom right, the symbols are: -

> ➤ *Dangerous to the environment*
> ➤ *Toxic*
> ➤ *Gas under pressure*
> ➤ *Corrosive*
> ➤ *Explosive*
> ➤ *Flammable*
> ➤ *Caution – used for less serious health hazards like skin irritation*
> ➤ *Oxidising*
> ➤ *Longer term health hazards such as carcinogenicity and respiratory sensitisation*

Controlled drugs - A controlled drug is one whose use and distribution is tightly controlled because of its abuse or risk. These drugs are rated in the order of their abuse risk and placed in grades. These drugs are placed in grades; the drugs with the highest abuse potential are placed in grade I, and those with the lowest abuse potential are in grades 3, 4 and 5, e.g.

Grade 1 Drugs	Grade 2 Drugs	Grades 3,4 and 5 Drugs
heroin, marijuana, LSD, PCP, and crack cocaine	morphine, cocaine, oxycodone methylphenidate and dextroamphetamin	Contain smaller amounts of certain narcotic and non-narcotic such as anti-anxiety drugs, tranquilizers, sedatives, stimulants, and non-narcotic analgesics, acetaminophen with codeine paregoric, hydrocodone with acetaminophen, diazepam (Valium), alprazolam (Xanax), propoxyphene (Darvon), and pentazocine (Talwin).

Controlled waste – Is waste that is subject to legislative control in either its handling or its disposal; the types of waste covered include domestic, commercial and industrial waste.

Convulsion - An involuntary contraction or series of contractions of the voluntary muscles; a type of Seizures consisting of a series of involuntary contractions of the voluntary muscles. Such seizures are symptomatic of some neurologic disorder; they are not in themselves a disease entity.

Coronary heart disease - *(CHD)* is a narrowing of the small blood vessels that supply blood and oxygen to the heart. CHD is also called coronary artery disease.

Cortisol - Known as 'the stress hormone; It is a hormone produced by the adrenal glands that helps to regulate the blood pressure and cardiovascular function, as well as the body's use of proteins, carbohydrates and fats.

Council – Body of people meeting to make decisions or laws

Cowper gland also called a bulbourethral gland - is one of two small exocrine glands in the reproductive system of many male mammals

Crack - Is the form of cocaine that can be smoked.

It may also be cavvy, **base**, or just **crack**; it is said to be the most addictive form of cocaine

Creutzfeldt–Jacob disease (CJD) - a degenerative neurological disease that is incurable and invariably fatal.

Sometimes called the human form of mad cow disease (***bovine spongiform encephalopathy or BSE).***

CJD is caused by an infectious agent called a Prion, progressive death of the brain's nerve cells.

CJD causes the brain tissue to degenerate rapidly, and as the disease destroys the brain, the brain develops holes and the texture changes to resemble that of a kitchen sponge.

Symptoms of CJD:

Rapid progression of dementia which leads to memory loss, personality changes, hallucinations, anxiety, depression, paranoia, obsessive-compulsive symptoms, and psychosis; accompanied by physical problems such as speech impairment, jerky movements (***myoclonus***), balance and coordination dysfunction (***ataxia***), posture becomes rigid, and frequent seizures occur

Crime prevention - The ultimate goal of preventing crime is to reduce the risk of being a victim. To effectively accomplish this, it is important to remove that opportunities a criminal will want to take advantage of you, your neighbour or property.

Criteria – A rule or principle for evaluating or testing something.

Crohn's disease – *(**Crohn's syndrome** and **regional enteritis**)* - is a type of inflammatory bowel disease **(IBD)** that may affect any part of the gastro-intestinal tract from mouth to anus; The three most common sites of intestinal involvement in Crohn's disease are ileal, ileocolic and colonic.

This is a lifelong inflammatory bowel disease (IBD). Parts of the digestive system get swollen and have deep sores called ulcers.

This disease is found in the last part of the small intestine and the first part of the large intestine.

It can develop anywhere in the digestive tract, from the mouth to the anus.

Cross-contamination - The physical movement or transfer of harmful bacteria from one person, object or place to another

Cross- Infection - The transmission of a pathogenic organism from one person to another: It is a common and important mode of infection with many varieties of organisms, including streptococcal and other bacterial diseases, viral hepatitis A and some other faecal-oral infections, such as scabies, fungus infections, pinworms, and roundworms

Cultural difference - The variations in the way of life, beliefs, traditions and laws between different countries, religions, societies and people.

Curtain – A hanging cloth (drape) on a window

Cut – An injury made by a sharp object (a cut on the skin)

Cycles of disadvantage – When someone is discriminated against, the negative results reflect and bring about chain reactions of more disadvantages

Cyst – A small blister-like structure formed in the body and could rupture unexpectedly (***Sebaceous cyst may also be associated with Gardner syndrome which consists of multiple colonic polyps, epidermoid cyst, fibroma and skull osteomas) see medical Journal –*** Not all cysts produce symptoms, pending upon the size and location on the body; if any signs or symptoms of a cyst is noticed, see a doctor who can diagnose the condition by examining the person concerned

Cystic fibrosis (*CF*) (*mucoviscidosis*) - is a recessive genetic disorder that affects mostly the lungs, pancreas, liver and intestine. A person with cystic Fibrosis develops Clubbing in the fingers of the hand.

*"D*iscretion will preserve you; understanding will keep you," (Proverbs 2:11)

D.S.M (Diagnostic and Statistical Manual of Mental Disorders) – It covers all mental health conditions for children and adults with details of all known probable causes and outcome

Daily Living Activities– The skills that affect directly, individuals in their everyday health and wellbeing

Dark – Without Light; pitch black

Data Protection – Data protection Act 1984

Data Protection Act 1998 – How personal information is acquired, maintained and stored

Day in; Day out – Every day's continuous actions for an indefinite period of time

Deaf – Someone that is not able to hear when spoken to

Death – The end of life and no more breathing

Death Erection (*terminal erection*) - is a post-mortem erection, technically a priapism, observed in the corpses of men who have been executed, particularly by hanging. Spinal cord injuries are known to be associated with priapism

Deceitful – The act of lying; being untruthful

Decision making - Is the cognitive process that results in the selection of a course of action among several alternative scenarios. Decision making process must produce a final outcome to which the output can be an action or an opinion of choice.

Decontamination -To make safe for unprotected personnel by removing, neutralizing, or destroying any harmful substance or poisonous gas

Deface – To disfigure; spoil in appearance

Degenerative – Conditions that get worse gradually instead of getting better

Dehydration - When the amount of water leaving the body is greater than the amount being taken in

Delusion - A belief held with strong conviction despite superior evidence to the contrary

Dementia – *(Latin, originally meaning "madness")* - The most potent and common form of Dementia Is Alzheimer's Disease. It is the loss of brain function that occurs with certain diseases affecting the memory, thinking, language, judgment, and behaviour: A large number of seniors suffer from Dementia. Dementia is caused due to damage of the connections between the brain and nerve endings. This ailment is not curable and its causes are unknown. There is no effective treatment for this disease, but some drugs may help to temporarily control the symptoms. The most potent and common form of Dementia Is Alzheimer's Disease

Demographic - A characteristic criterion used to classify people for statistical purposes, such as age, race, or gender.

Demonstrate - To prove or make clear by reasoning or evidence

Dental Erosion (*acid erosion*) - The irreversible loss of tooth structure due to chemical dissolution by acids not of bacterial origin; the damage caused by acids — often from food sources — softening the surface of the tooth's enamel, which is then more easily worn away.

Department of Works and Pensions (DWP) – The department responsible for the payments of benefits

Depression –Mental health condition where a person has a long lasting low mood, and loses interest in activities; a mood disorder. It causes persistent feeling of sadness and loss of interest. It can also be called major depressive disorder or clinical depression; it affects feeling, thinking and behaviour which can lead to varieties of emotional and physical problems.

Symptoms of depression - Fatigue or loss of energy almost on daily; excessive sleeping on every day basis, loss of interest in almost all activities (Anhedonia), restlessness (psychomotor agitation), retardation, gaining/losing weight. Depressive illnesses are disorders of the brain.

Dermatitis - Is inflammation of the skin. It has nothing to do with eczema; atopic dermatitis (AD) also known as atopic eczema is a type of eczema,

Dermatomyositis (*DM*) - is a connective-tissue disease related to polymyositis (PM) that is characterized by inflammation of the muscles and the skin; DM, frequently affects the skin and muscles, it is a systemic disorder that may also affect the joints, the oesophagus, the lungs, and, less commonly, the heart

Desk - Furniture used for writing and reading

Desomorphine - Commonly known as Crocodile drug (Krokodil in Russian) – A type of drug, synthesizable in an ordinary kitchen; the dirty Desomorphine contains toxic chemicals and users ignorantly inject the substance into their vein, destroying blood vessels and tissues. The street name **krokodil**) - is a derivative from morphine (***an opioid*** - natural and synthetic narcotics that have the same effects as opiates despite the fact that it is not obtained from opium) with powerful, fast-acting sedative and analgesic effects;

Develop – To grow into a more advance level

Develop - To bring into being gradually

Diabetes – A disease caused when there is a low blood sugar, glucose and insulin level; there are two types (type 1 and type 2).

When there is injury, attention must be paid to injury in a diabetic as the injury has a tendency to degenerate faster than a non-diabetic. It could result in the amputation of limbs; (this is common). Therefore, a quick diagnosis and treatment is very necessary for a Diabetic patient; a result of bad diet and considered to be a lifestyle disease. High fat and sugar intake may be the causative factors. Diabetes is caused due to the body's inability to produce insulin to keep blood sugar levels under control. This disease is curable with diet and medication either in injection or tablet form

Diabetic coma – Tiredness, the regular need to urinate and can slip into comma – *hospital treatment is recommended*

Diarrhoea – The frequent empting of the bowels with fluids in the faeces; There are three types of diarrhoea namely; **acute diarrhoea**, **persistent diarrhoea**, and **dysentery**. There are four clinical types of diarrhoea

> * **Acute watery diarrhoea** (cholera) lasts several hours or days: its main danger is dehydration; weight loss which may occur if feeding is not persistent.
> * **Acute bloody diarrhoea** (dysentery) its main dangers are intestinal damage, sepsis, malnutrition and complications which include dehydration that is likely to occur
> * **Persistent diarrhoea** (can last from 14 days or longer) its main danger is malnutrition and serious non-intestinal infection; dehydration may occur
> * **Diarrhoea with severe malnutrition** (kwashiorkor) its main dangers are: severe systemic infection, dehydration, heart failure and vitamin and mineral deficiency.

Die – When someone stops living; not breathing

Diet - Habitual nourishment; the kind and amount of food prescribed for a person or animal for a special reason

Digestion - The digestive system, also known as the **Gastrointestinal** Tract (GI) or the Alimentary Tract, is a long tube that is located between the mouth and the rectum. The digestive system is made up of several organs

Diphallus - It's when a man is born with two penises. A rare condition that affects one in 5-6 million males It's rare that both are fully functional, and it often comes in tandem with other deformities that also require surgery. Rare congenital anomaly in which the penis or clitoris is partially or completely doubled

Diploma – This is the certificate given to someone to confirm an honour or privilege

Diploma in Health and Social Care – A nationally recognized qualification by social care providers and the CQC (Care Quality Commission)

Diplomacy – The skill of negating; the art in handling people

Direct discrimination – Deliberate and obvious ill-treatment of someone or a group of people

Disable Person – Persons lacking in part or whole mobility to 100% mobility

Disability Discrimination Act 2005 – This Act protects people with substantial and on-going physical and/or mental impairment

Disablism *(Ageism)* – Someone being less properly looked after because they are old or vulnerable

Disclosure – Breaching of confidentiality in order to report information that may threaten the well-being of an individual.

Discriminate – Looking down on someone with less respect. (Discrimination Act 1995)

Discrimination – Unfair or Less favourable treatment of a person or group of people compared to the others with some kind of influence;

Discuss – Share with colleagues any experiences gained on confidential information with a service user or their relatives

Discussion – Sharing of ideas and exploring possibilities of improvement at work with colleagues

Disempowered – To put a limitation by curbing the powers of an individual in making decisions and to control staff

Dishonest – Being insincere; to cheat

Dismiss – To be unceremonially removed from a position of responsibility

Disrespect - To be rude; the lack of respect

Diversity – Variety; Different kind of people, able bodied and of all nationality; Being varied or different; the range of differences (age, gender, colour, cultural, etc.)

Documentation - Setting down on paper (writing down) experiences and observations as seen on a day- to- day basis, sharing and bringing fresh understanding

Door Entry System – Electronic device attached to an entrance door which allows it to be remotely controlled or Video Entry Controlled System

Doppler - *The Doppler Effect* holds true for all types of radiation, not just sound; the apparent change in frequency of sound wave echoes returning to a stationary source from a moving target. If the object is moving toward the source, the frequency increases; if the object is moving away, the frequency decreases.

Down syndrome - Occurs when an individual has a full or partial extra copy of chromosome 21; it was discovered in 1959 by the French physician Jérôme Lejeune; identifying Down syndrome as a chromosomal condition, instead of the usual 46 chromosomes present in each cell, he observed 47 in the cells of individuals with Down syndrome

Drugs –Medicine legally prescribed by a qualified a medical practitioner; any substance that alters normal bodily function, when absorbed into the body of a living organism. These alterations can be good or bad

Drug dependence - A compulsive need to use drugs in order to function normally; stopping the drug could lead to withdrawal symptoms. Drug addiction is the compulsive use of a substance, despite its negative or dangerous effects. A person may have physical dependence on a substance without being addicted e.g. certain blood pressure medications do not cause addiction but they can cause physical dependence.

Drug misuse – The habitual use of drugs to alter one's mood, emotion, or state of consciousness

Dukes' disease - known as **Filatov's disease**, is an exanthema - *an eruptive disease (as measles).* It is distinguished from measles or forms of rubella though it was considered as a form of viral rash - *A viral rash is usually a pink or red rash caused by a virus*

Duodenum - The first part of the small intestine. It is located between the stomach and the middle part of the small intestine; begins at the pyloric sphincter which receives partially digested food from the stomach and begins the absorption of nutrients. The duodenum is about 25 cm long, as seen in the diagram

Duty of care - is a legal obligation which is imposed on an individual that they comply to a standard of reasonable care while performing any acts that could cause harm to others; duty of care comes into force when one individual or group undertakes an activity which could reasonably harm another, either physically, mentally, or economically. It is also the legal obligation acting towards others with careful attention and reasonable caution in order to protect their wellbeing and harm occurring

Dwarf - (*see midget*) a dwarf is an extremely short adult who is less than 58 inches tall. The word midget is considered derogatory and should not be used to qualify such person(s)

Dysentery - Dysentery is very dangerous because of its ability to lead to anorexia, rapid weight loss, and damage to the intestinal mucosa; it is Diarrhoea *with blood in the stool* (with or without mucus)

Dyspareunia - (*Greek meaning "badly mated")* - This is a painful sexual intercourse, due to medical or psychological causes. The symptom is more common in women than in men. It affects one-fifth of women at some point in their lives. The causes are often reversible.

Dysrhythmia - cardiac dysrhythmia or **irregular heartbeat** are conditions in which the heartbeat is irregular too fast, or too slow.

"*E*xalt her, and she will promote you; She will bring you honour, when you embrace her." (Proverbs 4:8)

ECG - The *electrocardiogram* (ECG or EKG) is a diagnostic tool that measures and records the electrical activity of the heart in exquisite detail

Ear – That part of the head which we hear; it is made up of three different sections: the outer ear, the middle ear, and the inner ear. These parts all work together so you can hear and process sounds; also provide the sense of balance

Eating disorder - is a condition defined by abnormal eating habits. It involves insufficient intake of food to the disadvantage of an individual's physical and mental health *(see anorexia nervosa)*.

Ebola virus disease - *(EVD) or Ebola hemorrhagic fever (EHF)* - is a human disease caused by Ebola viruses. It is a fatal disease in humans and primates (such as monkeys) and also in bats. Fruit bats of the *Pteropodidae* family are considered to be the natural host of the Ebola virus; severely ill patients require intensive supportive care. No licensed specific treatment or vaccine is available for use in people or animals.

Eccentric – Not normal; irregular; odd behaviour; an individual who displays behaviour pattern that is not of the normal acceptable standards

Echoing - Is a reflection of sound, arriving at the listener sometime after the direct sound; the repetition of a sound by reflection of sound waves from a surface a sound so produced; the repetition of a sound by reflection of sound waves from a surface.

Ecstasy –Commonly known as *MDMA (3, 4-methylenedioxy-methamphetamine)*; a synthetic, psychoactive drug that has similarities to the stimulants *amphetamine* and *hallucinogen*

mescaline. It gives someone the feelings of increased energy, euphoria, emotional warmth and empathy toward others and distortions in sensory and time perception.

Eczema – (*atopic dermatitis*) medical conditions that cause the skin to become inflamed or irritated with itchy rash. There are many different types of eczema that produce symptoms and signs ranging from oozing blisters to crusty plaques of skin. Group of medical conditions that cause the skin to become inflamed or irritated

Edible – Fit to eat; fit to be eaten as food.

Effeminate – A man behaving like a woman; unmanly; having qualities and characteristics associated with women than men.

Effort – Hard work; a serious attempt

Effusion – (known as *joint effusion*) - is the presence of increased intra-articular fluid, it commonly affects the knee.

Ergonomics (*human factors*) - The scientific discipline concerned with the understanding of interactions among humans and other elements of a system, and the profession that applies theory, principles, data and methods to design in order to optimize human well-being and overall system performance.

Ergonomics: the science of designing user interaction with equipment and workplaces to fit the user (custom made designs; be it furniture, clothing, shoes, etc.

Ehlers–Danlos syndrome (EDS) - is an inherited connective tissue disorder with different presentations that have been classified into several primary types

Ehlers-Danlos syndrome type IV (EDS type IV) - is characterized by thin, translucent skin; easy bruising; characteristic facial appearance (in some individuals); and arterial, intestinal, and/or uterine fragility.

Eisenmenger's syndrome - Is a condition that affects blood flow from the heart to the lungs in some babies who have structural problems of the heart, it is caused by a defect in the heart. People with this condition are born with a hole between the two pumping chambers -- the left and right ventricles -- of the heart *(ventricular septal defect)*. The hole allows blood that has already picked up oxygen from the lungs to flow back into the lungs, instead of going out to the rest of the body.

Elder - Common tree of hedgerows. woods, chalk downs and waste ground, elder was once regarded as one of the most magically powerful of all plants; native to the British Isles. Commonly known as Tromán (Irish); Boon tree; Bore wood; Dog tree; Ellern or Fairy tree

Elderly (*older person*) – A senior person; older in age

Elderly Care Matters - *is* a free e-bulletin (electronic- bulletin) from Speech mark publishing, written by experts in the field of elderly care to make life as a professional easier

Electrocardiogram - *Electrocardiography* (**ECG** or **EKG**) is the interpretation of the electrical activities of the heart, across the thorax or chest over a period of time, as detected by electrodes attached to the surface of the skin and recorded by a device external to the body

Electronic Prescription Service (*EPS*) – A National Health Service (NHS) funded service in England, It gives you the opportunity to change how your General Practitioner (GP) or Doctor as they are known, will send your prescription to the professional healthcare chosen by you to get your medication from- *registered pharmacy.*

Emergency – Situation experiencing immediate action to be taken

Emergency Team – *(Telecare)* - Takes over from the manager when off site in sheltered Housing but on 24/7 with the vulnerable

Emotion – Mental state that arises spontaneously rather than through conscious effort and is often accompanied by physiological changes in which joy, sorrow, fear, hate, or the like, is experienced; mental reaction of anger or fear that is experienced as a strong feeling directed towards a specific object or person, accompanied by physiological and behavioural changes in the body

Emotional/psychological (*Abuse*) – Actions taken by another person that damages someone's mental wellbeing

Empathy – Understanding another person's feelings as if they are your own.

Empowerment –Giving power to; authorize, especially by legal or official means; more control over one's life through self-confidence and self- esteem; being involved in making decisions and overcoming obstacles which could include discrimination and/ or oppression

Enable - To make possible, practical, or easy

Enablement – Authority to do something

Encrypt – When e-mails, letters and files require password for access – "*computers*" **Endocrine system** - Is the system whereby each gland secretes different types of hormones directly into the bloodstream, some of which are transported along nerve tracts to maintain internal stability

Endometrial Cancer – (*Uterine Cancer)* - A type of cancer that begins in the inner lining of the uterus (*womb*); the fourth most common cancer in women

Endothelium - Is the thin layer of cells lining the interior surface of the blood vessels and lymphatic vessels forming an interface between circulating blood or lymph in the lumen and the rest of the vessel wall. The cells that form the endothelium are called *endothelial cells*

End of life care - refers to health care of all those with a terminal illness or terminal condition that has become advanced, progressive and incurable

Enlarged Prostrate - When the prostate gland located below the bladder in men produces fluid components of semen.

Over half of men aged 50plus have enlargement of the prostate gland. This condition is sometimes called benign prostatic hyperplasia or benign prostatic hypertrophy (BPH)

It is not known exactly why this enlargement occurs.

BPH is not cancer and does not cause cancer. Not all men have symptoms of BPH

Enter – To come in

Enthusiasm – Interest to assist in pushing forward

Entice – To tempt, arousing desire.

Entrust – Put into the care of another

Entrance – A place to enter, opening (door)

Environment – Natural surroundings

Enteritis – The inflammation of the intestines

Environmental Health Officer – Employed by the local government to protect public health. Their duties are to protect the working surroundings and prevent pollution

Envy – The feelings of uneasiness at someone else's well-being

Enzymes – Large biological molecules responsible for the thousands of chemical conversions that sustain life. Enzymes speed up chemical reactions in the body, but do not get used up in the process.

Ephelis (*Freckles*) - small brown spots usually found on the face and arms. Freckles are extremely common and are not a health threat. They are more often seen in the summer, especially among lighter-skinned people and people with light or red hair; people who freckle easily (for example, lighter-skinned people) are more likely to develop skin cancer.

Epicondyle – (*epi meaning "upon"; condyle meaning "knuckle" or "rounded particular area*) - A rounded projection at the end of a bone, located on or above a condyle and usually serving as a place of attachment for ligaments and tendons: a projection on the surface of a bone; often an area for muscle and tendon attachment.

Epilepsy – A brain disorder in which a person has repeated seizures (convulsions) over time.

Epileptic convulsions (*Seizure*) – Tonic-clonic seizure have two stages, namely**, *Stiffness in the person's body and Twitching of the person's arms and legs*** – consciousness is lost and some people will wet themselves while seizure usually lasts for three minutes or more

Epithelial Tissues - Is one of the four major tissue types in the body that acts as an interface between the body and the rest of the world

Equality – To be of the same – age, rank, etc.; treating a person fairly or in a way that ensures non-discrimination

Equality Act 2010 – This aims to protect and prevent previous legislation associated with race, gender, disability, age and discrimination in work, education, services, facilities, etc.; it states that all family members, service users and vulnerable tenants are fairly treated and are able to gain access and benefits from services and that health professionals and colleagues are treated equal

Equal Opportunity – the way in diversity recruitment; a condition for an agreement that all people should be treated similarly, unhampered by artificial barriers or prejudices

Equality of opportunity – The situation where everyone has equal chance to grow

Equal Opportunities Commission - The Commission is a corporate body, set up under the Equal Opportunities Act 2008. It comprises the Chairperson and three other members

Equal opportunity policies - Where workers enjoy protections under federal and state labour laws; employers bear the responsibility for complying with these laws.

Equal Pay Act Amended 1983 – removes sexist in the place of gathering

Equity - The value of an ownership interest in property, including shareholders' equity in a business as in finance

Erectile dysfunction (*ED*) - (impotence) sexual dysfunction characterized by the inability to develop or maintain an erection of the penis during sexual performance; when a man can't get an erection to have sex or can't keep an erection long enough to finish having sex. (It used to be called impotence).

Escalate – To quickly increase in motion

Estrogen – (Greek *οἶστρος* (oistros), meaning "gadfly" - General name for a group of hormone compounds; sexual passion or desire, and the suffix -*gen*, meaning "producer of - **Oestrogens or estrogens** - General term for female steroid sex hormones that are secreted by the ovary and responsible for typical female sexual characteristics group of compounds named for their importance in both menstrual and estrous reproductive cycles; the primary female sex hormones. It is the main sex hormone in women and is essential to the menstrual cycle.

Natural estrogens are steroid hormones, while some synthetic ones are non-steroidal; estrogen may also be used in males for treatment of prostate cancer.

Ethics - Greek" ethos" meaning "character", known as **moral philosophy**; it is a branch of philosophy that involves defending and recommending concepts of right and wrong conduct.

Ethnic groups – characteristic of people, especially group (ethnic group) sharing a common and distinctive culture, religion, language and belief

Ethnic minorities - A group of people who differ in race or colour or in national, religious, or cultural origin from the majority population—of the country in which they live

Euphoria: - the feeling of extreme happiness. It is not an everyday feeling; the most common causes of euphoric symptoms are drug intoxication or drug abuse. Mania is a related symptom (**see Mania**)

European Convention on Human Rights 1953 – Treaty to protect human rights and fundamental freedom in Europe; it is based on the United Nations Convention on Human Rights, drawn up after the 2nd World War – All citizens are protected and covered under this convention and have the right of appeal on decisions and rulings through the European Court of Human Rights

European Union - The (**EU**) is an economic and political union of **28 states** that are located in Europe. It operates through a system of supranational independent institutions and intergovernmental negotiated decisions by the member states.

Euthanasia - The practice of intentionally ending a life in order to relieve pain and suffering

Evaluate - To determine the significance; its worth by careful appraisal and study; to think about something in order to pass decision over it

Evaluation – The process of close examination and questioning to gain and judge performances, usually against an agreed standard

Evidence - Is anything presented in support of an assertion? This support may be strong or weak; something that furnishes proof: testimony; specifically.

Excrete – To discharge (waste matter) from the blood, tissues, or organs

Expectation - State of looking forward to or anticipating; an expectant mental attitude

Experience – An event gained through involvement in or exposure to a particular thing

Extra Care – Sheltered Housing – An independent living environment for vulnerable people with a twenty-four hours help facilities, seven days per week (24/7)

Extra Care Housing – A type of housing that offers help and support

Eyes – Organs that detect light and convert it into electro-chemical impulses in neurons. The simplest photoreceptor cells in conscious vision

Eye contact - Two people look at each other's eyes at the same time." In human beings, eye contact is a form of nonverbal communication between them

Eye lid - An eyelid is a thin fold of skin that covers and protects the eye, it has the thinnest skin of the whole body.

"Fearing the Lord is the beginning of moral knowledge but fools despise wisdom and instruction" (Proverbs 1:7)

Facet joint - (Zygapophysial joint) - The joint between the superior articular process of one vertebra and the inferior articular process of the vertebra directly above it. There are two facet joints in each spinal motion segment.

Facial expression - Positions of the muscles beneath the skin of the face. These movements convey the emotional state of an individual towards someone or thing

Facilitate – To encourage and enable

Facility – Something (machinery) that make things happen

Faeces – Excrement (solid) from the anus

Fair Employment – Where equal opportunity is given to Male and Female in any place.

Falls - Falls cause moderate to severe injuries, such as hip fractures and head injuries, and increase the risk of early death. It is a public health problem that is largely preventable; one in every three adults age 65 and over falls frequently

Family – Descendants of a common ancestor

Fail – To become unsuccessful

Faint – To become unconscious; lacking courage

Fake – A false appearance

Fall – To drop from a height; to collapse

False – Untrue; deceiving

Family – The descendants of one ancestor

Fan – An electronic rotating device that causes a current of air

Fanconi syndrome (also known as **Fanconi's syndrome**) is a disease which affects the kidney in which glucose, amino acids, uric acid, phosphate and bicarbonate are conceded as urine, instead of being reabsorbed. Fanconi syndrome affects the proximal tubule, which is the first part of the tubule to process fluid after it is filtered through the glomerulus (*is a knot of extremely tiny blood vessels involved in filtering the blood to form urine.*). It may be inherited, or caused by drugs or heavy metals - *Not to be confused with Fanconi anemia*

Fanconi Anaemia - (*FA*) - is a rare, inherited blood disorder that leads to bone marrow failure.

Farce – A silly sham

Fast – To be quick

Father – A male parent

Fatigue – Weakness caused by hard work; Fatigue (weariness, tiredness, exhaustion, lethargy) common causes include sleep problems, heart disease, lung disease, medications, endocrine disorders.

Favour – Goodwill; a kind of deed

Fear - Emotion aroused by danger

Feed – To give food to

Feedback – Written or spoken information about someone's performance from another person's point of observation

Fed-up – Tired and disgusted

Fence - Barrier used for enclosing land

Fever – Increase in the body's temperature above the normal set point.

Fibromyalgia – A condition that causes pain all over the body. It can affect anyone but commonly occur between 30 and 60years of age and can develop in people of any age, including children and the elderly. Symptoms can include anxiety, depression, fatigue, stiffness, headaches, Irritable Bowel Syndrome (IBS), and more; a disorder of unknown chronic condition that causes pain in the muscles, tendons and joints all over the body.

People with fibromyalgia have "tender points" on the body. *(See-Tender points)*; fibromyalgia also has other symptoms, such as -Trouble sleeping, Morning stiffness, Headaches, Painful menstrual periods, Tingling or numbness in hands and feet, Problems with thinking and memory (sometimes called "fibro fog") – *see fibro fog.*

No one really knows what causes fibromyalgia

Fibro fog - is a physical symptom of fibromyalgia, not a psychological one. While research on fibro fog is still going on, experts agree that it is not the result of Alzheimer's disease, dementia, or other brain function deterioration condition. The lack of information as to fibromyalgia's origin, surrounds fibro fog; variety of causes have been proposed, including depression, decreased oxygen flow to the brain, certain medications, poor nutrition, and changes to the Central Nervous System (CNS), most experts agree that sleep deprivation is the primary culprit of fibro fog.

Fibromyalgia Syndrome - is a common and chronic disorder characterized by widespread muscle pain, fatigue, and multiple tender points; for some unknown reasons, between 80 and 90 percent of those diagnosed with fibromyalgia are women; however, men and children also can be affected. Several studies indicate that women who have a family member with fibromyalgia are more likely to have fibromyalgia themselves, but the exact reason for this—whether it is heredity, shared environmental factors, or both—is unknown.

Fight – Physical aggressive combat by hand

Filatov's disease - known as **Dukes' disease**, is an exanthema - (*an eruptive disease (as measles).* It is distinguished from measles or forms of rubella though it was considered as a form of viral rash - *A viral rash is usually a pink or red rash caused by a virus*

File – A folder cabinet in which papers are kept

Financial (*Abuse*) – The misuse and theft of money from the service users and vulnerable tenants that are being supported

Finger – One of the five end part of the hand

Fire – Heat and light caused by destructive burning (of a house, bush, etc).

Fit – (as in sickness) - an attack of illness

Fit – Suitable – for...

Flabby – Loose hanging

Flag – A piece of cloth with design to show nationality; flat paving stone

Flexibility – easy to bend; usually without breaking

Flu - called "Influenza" is caused by viruses that infect the respiratory tract: Influenza viruses are divided into three types designated A, B, and C, with influenza A types usually causing the most problems in humans. (*See-Influenza*) Flu is not only unpleasant but can cause serious illnesses such as pneumonia or bronchitis and in the worst scenario, death.

Flu jab – Flu vaccine given to people with long term health condition or are aged 65years and over to help them stay healthy in winter

Fluid restriction - Another name for Congestive Heart Failure is Congestive Heart Failure. Some people with severe congestive heart failure may require fluid restriction

Floor – That part of a room that one stands on.

Fly – An insect with two transparent wings found in warm untidy areas

Fob – A flat square object electronically programmed to be used as a key for security purpose.

Follicular Unit Transplantation (FUT) - a technique which involves a thin strip of donor hair being transplanted from the area of permanent hair; strip is removed under local anesthetic and the area, which is usually at the back of the head, is sutured (*a stitch used by doctors and surgeons to hold tissue together*).

Food – Nourishment that we feed on

Food Safety – It is for Carers that are involved in preparation of food and drinks for clients.

Foolscap – Printing or writing paper

Foot – That part of the body that an animal stands and walks on; the end part of the leg

Formal relationship – Friendship between colleagues in the place of work

Fragile – Frail

Freckles (*Ephelis*) - small brown spots usually found on the face and arms. Freckles are extremely common and are not a health threat. They are more often seen in the summer, especially among lighter-skinned people and people with light or red hair; people who freckle easily (for example, lighter-skinned people) are more likely to develop skin cancer.

Freedom of Information Act 2000 – Protects the rights of individuals to have access to personal data held about them

Freeze – To become ice

Frenzy – A violent excitement

Frequent – Occurring often

Friend – A well-wisher; an acquaintance someone whom one knows

Functional skill – Type of qualifications developed by the UK Government as part of an initiative to improve the country's literacy, numeracy and ICT skills.

Funeral – A ceremony of cremation or burial

Fuss – Unnecessary complaints or protest

"Go to the ant, you sluggard; observe its ways and be wise!" (Proverbs 6:6)

Gadfly - A fly that annoys horses and other livestock, usually a horse-fly or a botfly - *Gadfly ethics was used by Plato in the Apology to describe Socrates's relationship of uncomfortable goad to the Athenian political scene-*

Gardner syndrome - is an autosomal dominant premalignant syndrome. *(See cyst)*

Gastric dilatation volvulus - *known as Bloat or twisted stomach* - It is a medical condition in which the stomach becomes overstretched by excessive gas content; commonly referred to as **torsion** and **gastric torsion** when the stomach is twisted. *Bloat* is often used as a general term to cover gas distension of the stomach with or without twisting.

Gastrin - a peptide hormone that stimulates secretion of gastric acid (HCl) by the parietal cells of the stomach and aids in gastric motility.

General Practitioner *(GP)* - is a medical practitioner –a medical doctor- who treats acute and chronic illnesses and provides preventive care and health education to patients. "The good GP is a medical doctor who will treat patients both as people and as a population"

Genital engorgement - For a man engorgement is his erection, when the penis becomes engorged with blood.

For a woman, it is her expanded, lengthened, ready vagina. Engorged Genitals occur during Pregnancy: it causes sexual arousal and intense orgasms often with no gratification no matter how many orgasms the woman may have. Due to increased blood flow through the whole body and by the second trimester, genitals become engorged due to pooling blood.

Gerontology -The branch of science that deals with aging and the problems of aged persons; the physical, mental, and sociological study of aging

Gerontology Geriatrics - A multidisciplinary field; meaning the study of aging combines or integrates information from several separate areas of study.

Gingivitis ("inflammation of the gum tissue") - A non-destructive periodontal disease

The most common form of gingivitis, and periodontal disease is called plaque - In the absence of treatment, gingivitis may progress to Periodontitis, which is a destructive form of periodontal disease.

Glaucoma - Glaucoma is the second leading cause of blindness; Glaucoma is actually a group of diseases. The most common type is hereditary. Glaucoma is a group of eye diseases that usually share common traits, such as high eye pressure, damage to the optic nerve and gradual sight loss. Most kinds of glaucoma involve elevated eye pressure.

Types of Glaucoma - Other types of glaucoma are variations of open-angle or angle-closure types. These types can occur in one or both of your eyes

Angle-Closure Glaucoma - (acute glaucoma or narrow angle glaucoma)	- It is much rarer and is very different from open-angle glaucoma in that the eye pressure usually rises very quickly. This happens when the drainage canals get blocked or covered over, like a sink with something covering the drain.
Congenital glaucoma - (Childhood glaucoma)	–The common term used for glaucoma diagnosed in early childhood It can come about by hereditary defect or abnormal development during pregnancy, Childhood glaucoma, referred to as congenital glaucoma, paediatric, or infantile glaucoma that occurs in babies and young children. It is usually diagnosed within the first year of life; a rare condition that may be inherited and caused by incorrect development of the eye's drainage system before birth.
Irido Corneal Endothelial Syndrome	- occurs more frequently in light-skinned females. Laser therapy is not effective in these cases. The symptoms include hazy vision upon awakening and the appearance of halos around lights. Its treatment includes medications and filtering surgery.
Neovascular glaucoma	- is always associated with abnormalities such as diabetes; very difficult to treat. It never occurs on its own.
Normal-Tension Glaucoma	– (Low-tension or normal-pressure glaucoma) - in normal-tension glaucoma the optic nerve is damaged even though the pressure in the eye is not very high. The causes of NTG still remain unknown
Pigmentary Glaucoma	- is a form of secondary open-angle glaucoma. When the pigment granules in the back of the iris (the coloured part of the eye) break into the clear fluid produced inside the eye.

Primary Open-Angle Glaucoma	- the most common form of glaucoma, affecting millions of people, it happens when the eye's drainage canals become clogged over time; the inner eye pressure *(intraocular pressure or IOP)* rises because the correct amount of fluid can't drain out of the eye.
Pseudoexfoliative Glaucoma	- Secondary open-angle glaucoma occurs when a flaky, dandruff-like material peels off the outer layer of the lens within the eye. Common among Scandinavian descendants'
Secondary glaucoma	– This is any case in which another disease causes an increased eye pressure, resulting in optic nerve damage and vision loss.
Traumatic glaucoma	- occur immediately after an injury to the eyes or in later years. It can be caused by blunt injuries that bruise the eye (called blunt trauma) or by injuries that penetrate the eye. Conditions such as severe near-sightedness, previous injury, infection or prior surgery may also make the eye more vulnerable to a serious eye injury; common cause is from sports-related injuries, such as baseball or boxing

Glucose-6-Phosphate Dehydrogenize (*G6PD*) - a metabolic enzyme involved in the pentose phosphate pathway, especially important in red blood cell metabolism. G6PD deficiency is the most common human enzyme defect. G6PD deficiency, which leads to decreased NADPH (***Nicotinamide adenine dinucleotide phosphate***) or in older notation, TPN (***Triphosphopyridine nucleotide***) is a coenzyme used in anabolic reactions) levels

Glycation (*non-enzymatic glycosylation*) - the covalent (*a chemical bond that involves the sharing of electron pairs between atoms*) bonding of a protein with a sugar molecule, such as fructose or glucose, without the controlling action of an enzyme; the reaction that takes place when simple sugar molecules, such as fructose or glucose, become attached to proteins or lipid fats without the control of an enzyme

Glycolysis - The metabolic breakdown of glucose and other sugars that releases energy in the form of ATP; the ability of a cell regulating its internal conditions usually by a system of feedback controls, so as to stabilize health and functioning, regardless of the outside changing conditions

Gout – Known as "**Podagra**" when involving the big toe. It is a medical condition usually known by its recurrent attacks of acute inflammatory arthritis; Gout is a kind of arthritis that occurs when uric acid builds up in blood and causes joint inflammation. Known as gouty arthritis, is incredibly common and painful. People with gout usually first experience pain in the joint of the big toe

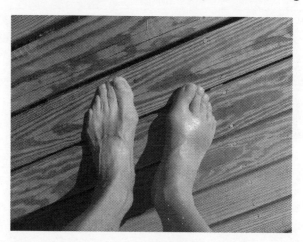

Governance – Checks and procedures used in monitoring, in order that the correct rules are followed

Gratuity – Gift of money in exchange for services rendered

Grave –a place where a dead body is buried in the ground and a stone memorial is placed

Gravity – Seriousness, as in an offence

Grey – As in having grey hair

Grievance – Ground made for complain

Grief – To mourn

Groundless – without reason

Ground floor – The floor on the ground level

Ground rent – Rent paid to the land lord, this included in the council tax

Grumpy – Always complaining and moaning

Guest – Visitor; an invited person who spends some time at another person's home in some social activity, as a visit, dinner, or party

Guarantor – Someone who promises to be responsible for another person's fulfilling promise

Guardian – Someone protecting the interests of another person; someone legally entrusted with the care of another or property.

Gutter – Pipe that channels water to its required source
Government – Persons appointed to rule a country or state.

Guided discovery – Words commonly used by therapists to assist people (individuals) think in alternative ways that could lead to changes in behaviours

*"H*onour the Lord from your wealth and from the first fruits of all your crops" (proverbs 3:9)

HRT – (*see Hormone Replacement Therapy*)

Habit – An unconscious pattern of behaviour that is acquired through frequent repetition

Haematoma – When the wall of the vein has been punctured and blood has leaked to tissues where it should not be.

Haemophilia – is a group of hereditary genetic disorders that impair the body's ability to control blood clotting: used to stop bleeding when a blood vessel is broken.

Half – brother/sister – The brother or sister by one parent only

Hallucinate – Vision that does not exist; to hallucinate is to see things that are not really there, or to distort objects that are there; a sensory perception with no outward stimuli e.g. hearing sounds

Hair dresser – Someone who washes and trims hair

Hair brush – Brush for the hair

Hand book - Hand held book containing important information

Handicapped – A physically or mentally disabled person

Handover – To relinquish authority to another person

Handyman – Odd job person; a person skilled at a wide range of repairs, typically around the home. These tasks include trade skills, repair work, maintenance work and both interior decorations

Hate – To dislike; Hatred (or hate) is a deep and emotional extreme dislike that can be directed against individuals, entities, objects, or ideas.

Hazard – Anything which might cause danger; a situation that poses a level of threat to life, health, property, or environment. An unavoidable danger or risk, even though often foreseeable

Hawker – Someone who goes about by offering something to sell

Hay fever - Seasonal irritation of the Nose, Throat, Eyes, etc. by pollen

Head – Uppermost part of the human body; the upper part of the body in humans, joined to the trunk by the neck, containing the brain, eyes, ears, nose, and mouth

Headboard - The panel at the head of the bed

Headstrong – Someone that is stubborn and self-willed

Heal – Restore; to be restored to good health

Health – The wellness, fitness and the general condition of a person's mind and body, usually being free from illness, injury or pain (as in *"good health"* or *"healthy"*)

Health education – An education to the general public that attempts to improve, maintain and safeguard the health care of the community.

Health education campaigns - One aspect of the many and larger programs of professional, patient, and public education designed to reduce the risk and consequences of heart, lung, and blood disease

Health and Social Care (HSC) Act2008: HSC Act (Regulated Activities) Regulations 2010: Care Quality Commission (CQC) Regulations 2009 – The CQC was established to regulate the Safety and Quality of health and social care services, replacing the National Minimum Standards (NMS)

Healthy – Someone in good health

Health Authorities – This is the Organisation that manages the health care e.g. NHS (National Health Services)

Health Minister – An appointed member of parliament, representing the government, who is responsible for the department of health and social welfare of the country

Health and Safety at Work Act 1974 – This Act recommends that all employees and their employers are to be made aware of Health and Safety issues and personal responsibilities; regular health and safety inspections are to be carried out by inspectors

Health and Safety Executive – The health and Safety Inspectorate is responsible to the executives when carrying out inspections of premises; it ensures that accidents are investigated and the implementation of "RIDDOR"- Reporting of Injuries, Diseases and Dangerous Occurrences; governmental department responsible for monitoring and making sure that the health and safety laws are carried out in the place of work

Health and Safety – Making sure that equipment and chemicals conform to the regulated procedure when being used and stored

Health & Social Care – A qualification which qualifies the worker 's competence in executing the roles assigned to them as carers or managers. There are courses relating to this in Levels 2, 3 and 5.

Health and Social Care worker – Someone with a health and social care qualification who works in the social care environment

Health and Safety Regulations – Management of Health and Safety at Work 1999 (*employers to train staff in health and safety regulations, fire, risks assessments, etc.*): The Manual Handling Operations 1992 (amended 2002) (*covers all manual handling activities, pulling or carrying objects and people*): H&S –First Aid 1981 (*providing first aid in the place of work to staff members*): Reporting of Injuries, Diseases and Dangerous Occurrences Regulations (RIDDOR)1995 (*employers to make known to the health and safety executive, range of occupational injuries, diseases* *and dangerous events):* Control of Substances Hazardous to Health (COSHH) 2002 *(employers to take the necessary precautions to ensure the hazardous substances are stored and used correctly)*

Health Care - The diagnosis, treatment, and prevention of disease, illness, injury, and other physical and mental impairments in human beings

Health Care Assistant – *(HCA)* People working with the elderly who depend on your training, knowledge and experience for quality care, encouragement, and support

Health Care Associated Infection (HCAI) – Infection that comes from a health setting environment

Health Centre – A building where nurses, doctors and other health workers carry out their professionalism on the various patients

Health targets - Government set targets to reach all 10 of its public-sector targets on welfare, crime, health and other areas.

Health Visitor – A medical professional who visits patients in their homes

Hearse – The vehicle or horses drawn carriage that carries a coffin; a funeral vehicle used to carry a coffin from a church or funeral home to a cemetery.

Heart – The most important part of the body that circulates the blood to keep someone alive

Heart-attack – *(myocardial infarction)* - the flow of oxygen-rich blood to a section of heart muscle suddenly becomes blocked; the chest feels as if it is being pressed by a heavy object, the pain radiates from the chest to the jaw, neck, arms and back, shortage of breath with an overwhelm anxiety

Heart-burn – The burning feeling in the chest caused by indigestion

Heart failure - Heart failure does not mean the heart has stopped working, it means that the heart is weak at pumping blood than normal; blood moves through the heart and body at a slower speed and pressure, as a result, the heart cannot pump enough oxygen to meet the needs of the body

Heart Rate – The rate at which the heart beats per minutes (70 times for an adult)

Heel – The back part of the feet

Hepatitis – The inflammations of the liver, caused by infection, damage, chemical misuse or drug treatment. The most common forms of hepatitis are: -

Hepatitis A – Transmitted by food or water contaminated by faeces when hands are not washed when preparing food

Hepatitis B – This is sexually transmitted to an unborn baby by the mother through blood transfusion or contaminated needles

Hepatitis C – Virus contaminated blood and needles – No treatment found as of now

Hepatitis E – Similar to hepatitis A which is transmitted through drinking water contaminated by faeces

Herbal – Medicine made of herbs

Herbalist – Someone who is involve in herbs

Hernia – A bodily organ through a weak spot in its surrounding walls

Heroin - Also known as **diamorphine** (BAN), and colloquially as **H**, **smack**, **horse**, **brown**, **black**, **tar** etc.; an active drug, converts into morphine in the body.

High Blood Pressure – This is when the blood pressure is above normal; it can be controlled by medication, kind of food we eat, exercise, etc.

Hoarder – Someone who stores items in great quantity and become a health hazard to the person

Hoist – Equipment used for lifting a disable person into a bath or while changing; it comes in

Electronic or manual forms

Holism – The whole person; taken into account their needs, wishes, situation, etc.

Holistic – Taking into account concerns about the complete person rather than with the analysis of the mind and the body, separately

Holistic model – The "whole person" by looking at the physical, mental, emotional and spiritual aspects of someone's health and wellbeing

Holistic Care – Taking care of the whole human body which involves the physical, social, religion, emotional and cultural needs are taken into account and not separately e.g. in a hospice

Hormone Replacement Therapy (*HRT*) - a common treatment for symptoms of menopause and aging

It is usually prescribed by Doctors during or after menopause; HRT in menopause is medical treatment in surgically menopausal, premenopausal (*covers symptoms, treatment, self-care of this life transition for women)* and postmenopausal (*medical information written for you to make informed decisions about your health concerns*) women.

HRT ease discomfort caused by less circulation of estrogen and progesterone hormones in menopause.

HRT occurs after a woman's period stops, her hormone levels fall, causing uncomfortable symptoms like hot flashes and vaginal dryness and sometimes conditions like osteoporosis. HRT replaces hormones your body no longer makes. It's the most effective treatment for menopause symptoms.

Homosexuality – The sexual preference for the same gender

Hose – A flexible pipe-like object used to convey liquid e.g. water, oil

Hospice – A place set up for taking care of the dying; providing counselling, holistic care and palliative care (total support for the dying)

Hospital – Place where the sick and injured are taken to for treatment

Host body – Someone who accommodates virus or parasite within them

Hot flashes (also known as **hot flushes**) are a symptom which may have several other causes, but which is often caused by the changing hormone levels that are characteristic of menopause. They are typically experienced as a feeling of intense heat with sweating and rapid heartbeat, and may typically last from two to thirty minutes for each occurrence. Hot flashes in men could have various causes. One is a possible sign of low testosterone (*see HRT)*

House-bound – Unable to leave their homes on their own due to ill health

House-proud – Keeping one's house or place clean and tidy

Housing – Places of shelter where people live

Houses of Parliament - Commonly known as the ***Houses of Parliament*** after its tenants *The Palace of Westminster* is the meeting place of the House of Commons and the House of Lords, the two houses of the Parliament of the United Kingdom.

Huntington's disease – This is a hereditary disease of the central nervous system; a common disease in the United Kingdom; affecting people directly or indirectly due to a faulty gene *(Chorea)*. **Huntington's disease** (*HD*) - is a neurodegenerative genetic disorder that affects muscle coordination and leads to cognitive decline and psychiatric problems. It typically becomes noticeable in mid-adult life.

Human rights - Based on the principle of respect for the individual; their fundamental assumption is that each person is a moral and rational being who deserves to be treated with dignity. They are called human rights because they are universal. Whereas nations or specialized groups enjoy specific rights that apply only to them, human rights are the rights to which everyone is entitled—no matter who they are or where they live—simply because they are alive.

Hump – This is a lump on the back

Hunger – The lack of food

Husband – The man to whom a woman is married

Hygiene – The practices of achieving cleanliness and maintaining good health; the procedure to maintain cleanliness as in health and safety

Hypercholesterolemia - is a condition characterized by very high levels of cholesterol in the blood. Cholesterol is a waxy, fat-like substance that is produced in the body and obtained from foods that come from animals (*particularly egg yolks, meat, poultry, fish, and dairy products);* the body needs this substance to build cell membranes however, too much cholesterol, increases a someone's risk of developing heart disease.

Hyperhomocysteinemia - Excessive homocysteine levels in blood. It is often associated with folate or cobalamin deficiency as well as genetic defects.

Hyperglycaemia (*high blood sugar*) -Is a condition in which an excessive amount of glucose circulates in the blood plasma.

Hypermetropia - Defect of vision in which a person is able to focus on objects in the distance, but not on close objects.

Hyperthyroidism – (Overactive thyroid) is a condition in which the thyroid gland produces too much thyroid hormone.

Hypo-allergenic - Meaning "below normal"; to have a decreased tendency to cause allergies; hypo means less, not none. Hypoallergenic is the characteristic of provoking fewer allergic reactions in allergy sufferers.

Hypoglycaemia (*low sugar blood*) –A medical emergency that involves an abnormally diminished content of glucose in the blood

Hypohydrosis - diminished sweating in response to appropriate stimuli. It can lead to hyperthermia, heat exhaustion, heat stroke and potentially death. An extreme case of hypohydrosis where there is a complete absence of sweating and the skin is dry is known as Anhidrosis.

Hypothermia – Loss of body heat below the normal body temperature commonly occur among vulnerable people e.g. elderly and babies

Hypotonia - the state of low muscle tone when the amount of tension or resistance to stretch in a muscle, often involve reduced muscle strength; it is a sign of worrisome problem. This condition can affect children or adults

"*I*f you are wise, you are wise to your own advantage, but if you are a mocker, you alone must bear it." (Proverbs 9:12)

I.C.D (*International Classification of Diseases*) – Published by World Health Organisation (*WHO*) identifying symptoms, findings and external causes of a range of mental illnesses

IV Cannulation - Intravenous Cannulation is a technique in which a ***cannula*** (a tube that can be inserted into the body, often for the delivery or removal of fluid) is placed inside a vein to provide venous access. Venous access allows sampling of blood as well as administration of fluids, medications, chemotherapy, and blood products

IV Medication - Intravenous (IV) medication administration refers to the process of giving medication directly into a patient's vein.

Ice – The solid form of water

Identity – A person's conception and expression of their individuality; the set of characteristics by which a thing or person is recognised

Ignorant – Lacking factual knowledge

Ileostomy - a surgical opening constructed by bringing the end or loop of small intestine (the ileum) out onto the surface of the skin.

Ill – Not well, sick

Illness – Unhealthy condition; poor health; sickness.

Ill-tempered - Having a bad temper; irritable

Immunisation - The process by which an individual's immune system becomes fortified against an agent (known as the immunogen); the most important way of protecting individuals and the community from vaccine preventable infectious diseases.

Immunity - Resistance of an organism to infection or disease; the natural defence of the body fights against the infection which is provided by the immune system

Impersonate – Pretend to be someone else:

Impinge – To have an effect upon the imagination

Implicate – To involve intimately: evidence that shows a connection with crime

Implement - To carry out; fulfil; perform

Impotent – Lacking in strength; unable to engage in sexual intercourse because of Erectile dysfunction;

Impose – To enforce a burden upon

Industrial Tribunals - Independent judicial bodies that hear and determine claims to do with employment matters; these include a range of claims relating to unfair dismissal, breach of contract, wages/other payments, as well as discrimination on the grounds of sex, race, disability, sexual orientation, age, part time working or equal pay

In Two Minds – Undecided; cannot make a decision

Incidence – The number of new cases of death or injury that develop during a specific period of time

Independent Complaints Advocacy Service (*ICAS*) –Independent and free service made available to users of health services, relatives and carers in the United Kingdom

Independent living – The opportunities of persons with disabilities to gain more personal power, self-determination and equality

Incinerate – *(Latin- incinerātus)* - to burn or reduce to ashes; cremate.

Inclusion – Being part of *a team*

Inclusive practice – A behaviour that shows every individual is accepted, included and tolerated equally

Incontinent - The involuntary excretion of bowel contents Lack of self-control

Income support – Extra money you may be able to get, if you have no income or you're working less than 16 hours a week

Income – The amount of money received during a period of time in exchange for services rendered

Incompetent – Without adequate knowledge; lacking qualities such as skills or abilities needed to do a job

Inconvenient – Not appropriate; awkward

Independent – Free from the control of others

Index – Something that guides or possibly to facilitate references e.g. The index of a book

Indigestion – Common digestive disorder; a feeling of discomfort in the upper abdomen after feeding

Indirect discrimination – Unfair treatment and actions that occur behind the person concerned

Individual – Each one separately

Induce – To bring about or stimulate the occurrence of something, such as labour; to increase the production of an enzyme or other protein

Infection – This is a condition where the body is invaded by parasites that are not normally present within the body; there are various types of infections and some of the most common ones are fungal infection, Vaginal infection, Bacterial infection, Viral infections.

Infected aneurysm - It is the growth of microorganisms (bacteria or fungi) in the vessel wall; infection arising within a pre-existing arteriosclerotic aneurysm

Infection hazard – Situations that is likely to spread bacteria (*pathogens*)

Infection prevention -

-Laws relating to Infection prevention and control

Health and Safety at Work Act 1974	*(Standards to prevent infection occurring & spreading)*
Public Health (Control of Disease) Act 1984	*(Standards for sanitation, water supply and disposal of rubbish)*
Food Safety Act 1990	*(Concerns food production and consumption)*
Environmental Protection Act 1990	*(Ensures safe management [handling, transfer, disposal] of controlled waste)*
Management of Health and Safety at Work	*(Introduced Risks Assessment)*

Infestation – To overrun in quantities, enough to be harmful

Inflammation – *(Latin "inflammo" - "I set alight, I ignite")* the process by which the body's immune system malfunctions; inflammation actually helps to heal damaged muscle tissue

Influenza, commonly known as "the **flu**" - an infectious disease of birds and mammals caused by RNA viruses of the family Orthomyxoviridae, the influenza viruses.

Influenza symptoms - are chills, fever, runny nose, sore throat, muscle pains, headache (often severe), coughing, weakness/fatigue and general discomfort. Although, often confused with other influenza-like illnesses, especially the common cold, influenza is a more severe disease

Flu Vaccine – *(Influenza)* - is a serious disease that can lead to hospitalization and sometimes even death. Every flu season is different, and influenza infection can affect people differently. Even healthy people can get very sick from the flu and spread it to others. *(Over a period of 31 seasons between 1976 and 2007, estimates of flu-associated deaths - (http://www.cdc.gov/flu/about/disease/us_flu-related_deaths.htm) - in the United States range from a low of about 3,000 to a high of about 49,000 people. During a regular flu season, about 90 percent of deaths occur in people 65 years and older. "Flu season" in the United States can begin as early as October and last as late as May).*

Information - In its general sense, is "Knowledge communicated and received concerning particular facts or circumstances"; it cannot be predicted

Information Commissioners Office (ICO) – An independent authority set up to ensure that organisations understand their responsibilities and the Rights of the Public

Inject - Able to force a fluid into (as to inject a drug for medical purpose)

Innermost – Farthest inwards as inner chambers

Inoculation – The introduction of serum or fluid into tissues in the form of a vaccine

Insulin – This is a peptide hormone (substances made of amino acids that are secreted by the endocrine system); a drug used for the treatment of type 1 and type 2 diabetes mellitus.

In-service training – Training provided by employers to their staff in order to help them develop the skills needed for their various jobs

Inter action - Reciprocal action, effect, or influence; the direct effect that one has on another

Intercavernosal injection therapy is a medication injected directly into the penis to treat ED

Intercom – Telephone system within a building; intercommunication device, talkback or door phone. It is a stand-alone voice communications system for use within a building or buildings

Interpret - Explaining the meaning of; making sense of

Intraurethral therapy is a suppository medication that is inserted into the tip of the penis to treat erectile dysfunction - *ED*

Introduce – To present formally, a person to another so as to make acquainted to *(two or more persons)*

Institutional (*Abuse*) – The abuse and maltreatment of a person by the people within that institution; the maltreatment of a person (often children or older adults) from a system of power

Institutionalised ageism – Where older people are not given equal access to treatments and investigations that will improve their health; instead people under the age of sixty-five (65yrs) are given precedence

Institutional discrimination – The unequal treatment of any group through practices or policies that is operating within any organisation

Intrude – To enter without permission

Interpersonal skills – Used by a person to properly interact with others especially In the place of work

Interview – Formal meeting between a candidate and a representative for position

Intestine - The segment of the alimentary canal extending from the pyloric sphincter of the stomach to the anus; the small intestine (or small bowel) and the large intestine

Investigation – A careful look into an inquiry

Iodine – A chemical element found in some foods; the body needs iodine to make thyroid hormones.

Investigate – Using the library or internet resources to find out as much as someone can about a person or subject

Invite – To ask to come (to visit or stay)

Involve – Engaging as a participant

Ischemia - Is a restriction in blood supply to tissues, causing a shortage of oxygen and glucose needed for cellular metabolism (to keep tissue alive). Ischemia is generally caused by problems with blood vessels, with resultant damage to tissue. It also means *local anaemia* in a given part of a body sometimes resulting from congestion such as thrombosis

Isolate - Separate so as to be alone; quarantine, to keep (*an infected person*) from contactwith noninfected persons

Itch – Irritation on the skin that makes someone want to scratch

Ivy – Creeping evergreen plant that grows on walls and trees

"Judgments are prepared for scorners, and floggings for the backs of fools." (Proverbs 19:29)

Jab – Sudden thrust (slang word commonly used for vaccine)

Jacket – A small coat worn over clothing to keep someone warm or presentable

Janitor – Care-taker of a sheltered scheme

Jar – A glass bottle with a wide mouth; a rigid, cylindrical container with a wide opening usually made of glass or plastic. They are used for things that are relatively thick or viscous

Jaundice – A disease that makes the eyes and skin to turn yellow commonly affect new born babies

Jaw – Sections of the mouth in which the teeth are set; the jaws are bony and oppose vertically, comprising an *upper jaw* and a *lower jaw*. The **jaw** is any articulated structure at the entrance of the mouth used for grasping food

Jealous – To envy; wanting what belongs to another

Jest – A joke; something that makes one to laugh

Jobless – Someone with no paid work

Joint – The movable points where the bones of the skeleton meet

Joint Effusion - A **joint effusion** is the presence of increased intra-articular fluid in the knee; it affects any joint in the body.

Joint Commissioning – When two or more Agencies coordinate their efforts to take joint responsibility where care is provided

Journal – A book where daily activities are recorded (a diary)

Joyful – Showing gladness

Jubilee – A fifty (50) year's celebration of an occasion

Jug – This is a dish with handle and spout for pouring out liquid e.g. gravy

Juice – The liquid part of fruits

Juke-box – This is an instrumental box that plays and records automatically (can rarely be seen today due to modern technology)

July – The seventh month of the year

Jumble Sale – The sale of old clothing and other odd items

Jump – To jump by leaping suddenly

Jump suit – A lady's all in one suit

Jumper – A knitted throw-over loose fitting to keep warm

Junk – General rubbish

June – The sixth month of the year

Junkie – (slang) someone addicted to drugs

Justice – The right to fairness under the law

"Kings reign by means of me, and potentates decree righteousness" (Proverbs 8:15)

Kettle – Pot used for boiling liquid with spout, handle and cover

Key - A device by which something is screwed or turned; a part of a musical instrument that makes a definite kind of sound

Key Hole – The hole in which a key is placed

Kick – To hit with the foot in sudden violence

Kidney – The bean shaped organs in the lower back region of the body (usually a pair); main organ of the urinary system, it separate waste products from the body fluids

This is a pair of organs located in the back of the abdomen. Each one is about 4 or 5 inches in length and the size of a fist; the function is to filter the blood. All the blood in our bodies travel through the kidneys several times a day Kidneys remove wastes, control the body's fluid balance, and regulate the balance of electrolytes. As the kidneys filter blood, they create urine, which collects in the kidneys' pelvis -- funnel-shaped structures that drain down tubes called ureters to the bladder.

Kidney transplantation - Is the organ transplant of a kidney into a patient with end-stage renal disease.

Kindred - Related

Kitchen – The place in which food is prepared

Kleptomania - The violent urge to steal

Kleptomaniac – Someone who suffers from the violent urge to steal from another

Knead – To press together (like flour into dough)

Knee – The joint between the thigh and the chin bone

Knife – A metal instrument sharpened to cut

Knock – Clanking noise against a door

Knot –A lump made by twisting ends together and drawing tight the loops to obtain

Know – To be made aware of

Knowledge - skills acquired through experience or education; information which is correct is knowledge. Knowledge can always be supported by evidence.

Knuckle – The joints of a finger; one of the joints connecting the fingers to the hand.

Krebs cycle - is a part of cellular respiration. It is a series of chemical reactions used by all aerobic organisms to generate energy; The Krebs cycle refers to a complex series of chemical reactions that produce carbon dioxide and Adenosine triphosphate (ATP), some compound rich in energy

Krokodil – *see desomorphine* - This extremely addictive inject able opioid is called krokodil (pronounced like crocodile) or desomorphine. It's so named in part because users report black or green scaly skin as a side effect

"Long life is in her right hand; in her left hand are riches and honour."(Proverbs 3:16)

Labelling – Derogatory terms used in identifying types of behaviours or persons e.g." the mad man", "the miserable woman", "the nagging woman", "the grumpy man"

Labile diabetes - A type of diabetes when a person's blood sugar level often swings quickly from high to low and from low to high. Also known as "unstable diabetes" or "brittle diabetes"

Lack – To have none of; the insufficiency of funds

Lactation - The secretion of milk from the mammary glands and the period of time that a mother lactates to feed her young

The process occurs in all female mammals, although it predates mammals.

In humans the process of feeding milk is called *breastfeeding* or *nursing*.

Lacto- trope cell - is a cell in the anterior pituitary which produces Prolactin in response to hormonal signals including dopamine which is inhibitory and thyrotrophin-releasing hormone which is Secretagogue

Lady – The madam of a house; a woman with extremely good manners

Lame – To be disabled in one leg

Langerhans cells - dendritic cells (antigen-presenting immune cells) of the skin and mucosa, and contain large organelles called Birbeck granules. They are present in all layers of the epidermis, but are most prominent in the stratum spinosum

Language difference - Distinctions between language differences; key learning indicators that a child has a disability impeding communication.

Lard – The dissolved fat of a pig

Large Intestine - The large intestine absorbs again water then gets rid of the drier residues as faeces. The different sections of the large intestine are the ascending colon, transverse colon, descending colon, then sigmoid colon which leads to the anus. The anus has voluntary and involuntary sphincters and also has the ability to distinguish whether contents are gas or solid.

Larynx – The entry part of the wind pipe

Lasagne – Food made with flat pieces of pasta baked with cheese, tomato and meat

Laser Treatment – A method of treatment involving a thin ray of light mainly used for certain conditions in the eye or in some instances, cancer.

Lass – Cockney name for a girl

Latch – The small catch of a wood or metal to fasten a door

Late – Recently passed-on; dead; deceased

Laugh – making a joyous sound of happiness

Lawn – An area with short grass well looked after and preserved

Lazy – Someone not wanting to work; idle person not making any effort to help themselves

Leader – Someone in authority who leads by actions

Leading question – The question readily containing the answer to the question that is being asked

Leak – A hole through which liquid may pass e.g. hole in a bucket

Leather – The refined skin of an animal

Leave – Permission to be absent e.g. request of absenteeism

Learning Disabilities – People with limited mental capacity but can still function normally – causes can be birth injuries e.g. cerebral palsy; infection to the foetus in a woman's womb e.g. loss of mental functions; genetic disorders e.g. William syndrome

Learning opportunities – Opportunities in providing training, employment services and community skills

Leg – The limb that man walks on

Legal powers – These are powers that can intervene to enable abusers to be prosecuted e.g. offences against the Person Act 1861 (*physical abuse*); Sexual Offences Act 2003 (*sexual abuse*); Protection from Harassment Act 1997 (*Psychological abuse*) and Section 47 of the National Assistance Act 1948 (*neglect*)

Legislation – Laws made by Parliament which determines the different legal rights of individuals and organisations e.g. Community Care Act 1990, a policy framework for Health and Social Care organisations, clients, patients, carers and professionals; written laws by the Acts of Parliament; used to govern the use of information

Legible – Readable; capable of being read

Leisure – Ways of using free time e.g. watching TV, shopping, sports activities, etc.

Lungs – The respiratory organ of the body; the organs of breathing usually two sacks in the body

Leprosy – A contagious skin disease; a tropical disease mainly affecting the skin and nerves that can cause tissue change and, in severe cases, loss of sensation and disfigurement; leprosy is a skin infection caused by a ***mycobacterium Leprae***. It is transmitted following close personal contact and has an incubation period of 1-30 years. It can now be cured if treated with a combination of drugs.

Lesbian – Females that are attracted to another female

Lie – To be untruthful

Liar – Someone who tells lies; being untruthful

Life – The period between birth and death

Life Line – Line of communication; rope for support in times of rescue operations

Life Skills Training – Methods used in supporting those with disabilities allowing them to lead independent life-style

Life-style – The choice of how someone chooses to live their life; Healthy lifestyle is encouraged in order to live longer through diet, stress management, non-smoking, exercise, etc.

Lift – An electronic box-like metal case fitted in high rise building to convey people up and down; used in hospitals to move patients from one floor to the next while they are lying down or wheel chair with ease

Ligament – is a sheet or band of tough fibrous tissue that connects bones or cartilages at a joint or supports an organ, muscle, or other body parts; it is the fibrous tissues that connect bones to bones and is known as *articular ligament, articular larua fibrous ligament*. Ligaments are similar to tendons and fasciae as they are all made of collagen except that ***ligaments join one bone to another bone, tendons join muscle to bone and fasciae connect muscles to other muscles***. These are all found in the skeletal system of the human body.

Light - Something that makes things visible or affords illumination: a source as the lamp

Linen – Sheets of cloth; fabric woven from fibres mostly used as bed covers and table cloth

Listen – To pay attention; to give ear to

Limb - The upper and lower limbs are commonly called the arms and the legs in the human beings; most other mammals walk and run on all four limbs. In man, the arms are weaker, but very mobile allowing for a suitable reach at a distance ending in a specialized way capable of grasping and manipulating objects.

Linen – Sheets of cloth and tablecloth; clothes, table coverings, undergarments, or bedclothes made from cotton

Live – Having life; being capable of vital functions

Liver -a glandular vascular organ in vertebrates that secretes bile, stores and filters blood, and takes part in many metabolic functions such as the conversion of sugars into glycogen. The liver is reddish-brown and in humans is located in the upper right part of the abdominal cavity.

An internal organ of the body, protected by the ribs, dark-brown in colour It is the largest internal organ of the human body

Local – Confined to a place; pertaining to

Local authority – A body of people elected as local government; the body that has political and administrative powers to control a particular city or region in a country

Locker – A type of cupboard found in dressing rooms; used for containing temporary personal belonging during sports activities

Lock-Out – The act of locking out; tenants accidentally locked out by the wind when it blew the door shut

Lone – Companionless; lonely and having no companions

Louse – *(plural – Lice)*, a small wingless insect that lives as a parasite on humans and other animals. There are sucking lice, e.g. head and body lice, and biting lice, e.g. bird lice.

Lounge – A sitting room where someone can relax

Lowe Syndrome (LS) - (*see - Oculo-cerebro-renal (OCRL) syndrome*) is a rare genetic condition that causes physical and mental handicaps, and medical problems. It is also called the oculo-cerebro-renal (OCRL) syndrome: named after its discoverer, Dr. Charles Lowe and colleagues. Lowe syndrome can be considered a cause of Fanconi syndrome

Lupus - when the immune system attacks its tissues, causing inflammation, swelling, pain, and damage. Its symptoms include fatigue, joint pain, fever, and lupus rash; Lupus is an autoimmune disease that causes joint and organ damage.

Lurk – To hide and wait

Lust – A longing desire for power

Lump – A swelling; an abnormal swelling of the neck or leg

Lyme disease

Lyme disease is caused by the bacterium **Borrelia burgdorferi.**

It is send from one person, thing, or place to other humans through the bite of infected blacklegged ticks.

The symptoms include fever, headache, fatigue, and a characteristic skin rash called "chronic migrating redness" - *erythema migrans* - refers to the rash often seen in the early stage of Lyme disease.

The infection can spread to joints, the heart, and the nervous system. Currently, there is no human vaccine for Lyme disease.

"My child, if sinners try to entice you do not consent! (Proverbs 1:10)"

Macaroni – Wheat flour made in thick paste, then pressed through small holes and hardened, a well-known type of food

Mackintosh – Commonly called "Mac"; a waterproof covering or coat

Macaroon – Biscuit made of almonds

Mad – Insane

Madam – A way of addressing a lady

Mademoiselle – The French way of addressing a young lady

Madeira cake – Type of sponge cake

Magazine – Publication available at intervals (Weekly, Monthly or yearly)

Maggot - The legless, soft-bodied, wormlike larva of any of various flies, often found in decayed matter

Magnesia – Magnesium salt, white in colour used as medicine

Magnetic Resonance Imaging - Magnetic resonance imaging (MRI) is a test that uses a magnetic field and pulses of radio wave energy to make pictures of organs and structures inside the body

Mail – Parcels and letters sent by post

Maintenance – A means of support; to keep in good condition

Majority – Greater in number

Makaton – Signs and symbols used alongside speech to someone with learning difficulty to communicate

Malaria – A type of fever; a disease transmitted by the bite of an infected mosquito

Malfunction – Not working well

Malice – Spite

Mallet – a wooden hammer

Malnutrition - Under feeding; the condition that results from eating a diet in which nutrients are lacking; "malnourishment" is sometimes used instead of "malnutrition" Malnutrition is present in the form of under-nutrition, which is caused by a diet lacking nutrient food, and of poor quality.

Mammary Gland - an organ in female mammals that produces milk to feed young offspring; found in both sexes, but cease development in males well before puberty.

Mammography – The screening of the breast; using low dosage of x-ray to detect tumours or cysts

Manage – To deal with successfully; to be in charge of

Manager – Someone who can controllably carry out a task(s)

Mandatory – Compulsory; required

Mania is the mood of an abnormally elevated arousal energy level; a state of heightened overall activation with enhanced affective expression together with labiality of affect.

The signs and symptoms of mania -

- Elevated mood;
- Hyperactivity ;
- Excitement ;

- Overconfidence;
- Extravagance

Manual Pill Dispenser (*MPD*) – Sometimes called Dosett Boxes; they do not actually dispense pills. They provide specific sections for storage of drug doses; particularly useful for someone with short term memory problems where a carer or other individual can fill the containers with the appropriate doses. Contains four compartments in each tray labelled morn, noon, eve and bed in words or Braille

Mar-Chart (MAR referred to as **drug charts)** – It is a **Medication Administration Record** (MAR**) or (e-MAR** for electronic versions) a legal recording of the drugs that is administered to a patient at a hospital or home by a health care professional. The MAR is a part of a patient's permanent record in their medical chart. The health care professional signs off on the record at the time that the drug is administered. The electronic versions may be referred to as e-MARs.

Marfan syndrome - Mar fan syndrome is a heritable condition that affects the connective tissue; it affects different people in different ways.

Margarine – A butter-like vegetable fat used for cooking or spreading

Marigold – Type of plant related to daisy with yellow flowers

Marijuana – Dried flowers and leaves used as cigarettes for temporary exciting effects; expensive to purchase and causes damages to the brain tissues

Mascara – Colourings for the eyelashes commonly used by women and make-up artists

Mastectomy – The surgical removal of the breast commonly due to cancer

Masturbate – The stimulation of the sexual organ by use of the hand or other instruments; the sexual stimulation of one's own genitals, usually to the point of orgasm. Mutual masturbation can be a substitute for sexual penetration.

Mate – An equal or a companion

Matron – The woman in charge of nursing in hospital

May – The fifth month of the year

Meal – Food prepared and eaten

Meals On Wheels – This is a home service delivery of food in their homes (Elderly services)

Memories – The mental faculty of recalling past experiences based on the mental process of retaining and recognising information

Measles – An infectious red spots on the skin

Meat – Animal flesh used as food

Media – Ways of transmitting or conveying information e.g. recording or printing

Medical Care – Care given or administered medically by a health professional

Medical model of disability – The model by which disable people are defined by their illness, condition and diagnosis

Medicine – Substance used to keep away disease; the science of curing people

Meeting – Gathering; the act of coming together; an assembly of people or conference of persons for a specific purpose

Melanocytes are melanin-producing cells located in the bottom layer (stratum basale) of the skin's epidermis, the middle layer of the eye (the uvea), the inner ear, meninges, bones, and heart Melanin is the pigment primarily responsible for skin colour.

Melanoma - *Greek -melas*, "dark" - a type of skin cancer which forms from melanocytes (*pigment-containing cells in the skin*)

The most common site in women are the legs, and in men, the back. Particularly common among Caucasians, especially northern Europeans and north-western Europeans, living in sunny climates

Melasma (*Chloasma*) - A condition where tan or brown patches appear on the face

Melasma is a common skin disorder seen in men and especially women; commonly seen in pregnant women and often referred to as the mask of pregnancy.

Melasma frequently goes away after pregnancy.

This condition is worldwide and noticeable in all cultures and ethnicities; more common among Asians, Hispanics, Arabs and North Africans because of their higher levels of melanin in their skin Melasma is a harmless skin condition and not related to any medical disorder

Memoir – A written event documented from personal knowledge

Memory – The part of the mind where facts and experiences are stored and later remembered

Meningitis – Inflammations of the brain lining caused by bacteria

Menopause - the state of an absence of menstrual periods for a period of 12 months. The menopausal transition starts with varying menstrual cycle length and ends with the final menstrual period; Menopause is the time in a woman's life when the function of the ovaries ceases.

The ovary (female gonad), is one of a pair of reproductive glands in women.

They are located in the pelvis, one on each side of the uterus.

Each ovary is about the size and shape of an almond.

The ovaries produce eggs (ova) and female hormones such as estrogen, during each monthly menstrual cycle; an egg is released from one ovary and travels from the ovary through a Fallopian tube to the uterus.

Mental Health – Associated with someone's social and emotional state

Mental Health Disorder – The condition which affects the mental health of a person such as Learning disabilities, incomplete development of the mind, autism, dementia, etc.

Mental illness – The medical term used to qualify disorders of the human mind

Mental – Suffering from an illness of the mind

Menthol – A sharp smell from peppermint oil; used to relieve cold

Mentor – The giver of advice

Message – Information spoken or written passed from one person to another

Metastatic cancer - is a cancer that has spread from the part of the body where it started (the primary site) to other parts of the body; the spread of a cancer from one organ or part to another non-adjacent organ or part.

Meter – Instrument used for measuring e.g. electricity, water, etc. usage

Methods of Challenging Discrimination – There are four commonly known methods of challenging discrimination – ***Direct confrontation*** (an immediate response that gives a clear message of non-tolerance); ***Reporting discrimination*** (complaint procedures, used to record incidence of discrimination which should be followed up); ***Initiate discussion*** (sharing one's thoughts in a respectful manner with the person accused to reach an understanding); ***Support for those discriminated against*** (acting with confidence, without being confrontational and standing beside those on the receiving end of being discriminated at in a supportive action)

Methicillin Resistant - Staphylococcus aureus (MRSA) is a bacterium responsible for several difficult-to-treat infections in humans.

Methylated spirit – An alcohol based substance, unsuitable for drinking, used in hospitals for treatment

Micrograph - A drawing or photographic reproduction of an object as viewed through a microscope; digital image taken through a microscope or similar device to show a magnified image of an item.

Microwave oven - A gadget used for cooking and warming of food; produces heat by passing microwaves through the food

Midday – Middle of the day

Midnight – Middle of the night

Midget – *(see- dwarf)* anything very small of its kind; a person of extremely small stature who is otherwise normally proportioned.

Mind – The power of intelligence

Mirror – Reflective glass that gives a true picture of oneself

Mirroring - The behaviour in which one person copies another person while with them; it include miming gestures, movements, body language, muscle tensions, expressions, tones, eye movements, breathing, tempo, accent, attitude, choice of words or metaphors, and other aspects of communication. It is often observed among couples or close friends.

Mischief – Behaviour that causes annoyance to others

Misconduct – To behave badly

Miserable – To be unhappy

Mission – Tasks or operations that one conceives to carry out

Mistake – To make an error of (judgment)

Mistress – A woman with close sexual relationship with a man without being married to him

Mitochondria - are rod-shaped organelles that can be considered the power generators of the cell, converting oxygen and nutrients into adenosine triphosphate (**ATP**).

Mitten – A kind of gloves having one cover for all four fingers commonly used for babies

Mixed Blessing – Someone having both advantages and disadvantages over something

Mixed Marriage – Marriage between persons of different races religion or culture

Mock – To made a joke of; to laugh at

Modem – An electronic device which transmits information from one computer to another through a telephone line

Modest – Showing good behaviour

Modify – To alter

Moles - growths on the skin that are usually brown or black. Moles can appear anywhere on the skin, alone or in groups. Moles appear in early childhood and during the first 30 years of a person's life. A normal adult will have between 10-40 moles by adulthood. As the years pass by, moles change slowly, becoming raised and/

or changing colour. Hairs develop on the mole. Sometimes no change at all, while others may slowly disappear over time.

Monday – The second day of the week

Money – Printed notes and coins used in trading

Monitor – To watch over something; a screen used for transmitting images

Monogamy – Marriage to one wife or husband

Moon (*Braille*) – The methods by which sightless person can read using tactile dots or shapes; system which enables blind and partially sighted people to read and write through touch

Moral issues – The decision of what is right and what is wrong or appropriate in someone's behaviour

Morgue – A place where dead bodies are kept for identification

Mortuary – A place where dead bodies are kept before funeral or cremation

Mother – Female parent; a female head of a religious group (convent)

Mother Tongue – Someone's native language

Motivate - Acting towards a desired goal, controls, and sustaining certain goal-directed behaviours. It can be considered a driving force in someone's behaviour

Mourn – To grieve for e.g. a dead relative

Mouse – A small animal, very destructive and of many species

Mucosa - The moist tissue that lines certain parts of the inside of your body; it is found in the nose, mouth, lungs, and the urinary and digestive tracts.

Mucus – A slimy fluid from the nose produced by the mucous membrane

Multi- agency – Working together by two or more organisations to achieve a common goal

Multi-disciplinary – A combined team from different professional organisation for a cause

Multiple myeloma (*Greek myelo- meaning marrow*), known as **plasma cell myeloma** - A type of white blood cell that is normally responsible for producing antibodies.

Multiple Sclerosis (*MS*) – A disease of the nervous system; affects women more than men. MS is a chronic, often disabling disease that attacks the central nervous system (CNS), which is made up of the brain, spinal cord, and optic nerves; a disease of the nervous system which causes degenerative change to the spinal cord, optic nerves and the brain; a serious progressive disease of the central nervous system, occurring mainly in young adults and thought to be caused by a malfunction of the immune system. It leads to the loss of myelin in the brain or spinal cord and causes muscle weakness, poor eyesight, slow speech, and some inability to move **Multiple sclerosis (*MS*) (*disseminated sclerosis / encephalomyelitis disseminate*)** - is an inflammatory disease in which damages are caused to the insulating covers of nerve cells in the brain and spinal cord.

Muscles – Bundle of fibres in the body which causes movement

Muscular-Dystrophy - a medical condition in which there is gradual wasting and weakening of the skeletal muscles; **Muscular dystrophy** (**MD**) - is a group of muscle diseases that weaken the musculoskeletal system and impede locomotion.

Muscular tissue – There are *three types* of muscular tissues; *striped muscles* (voluntary muscles tissue); *Cardiac muscles* (involuntary muscles tissue); *Smooth muscle* (involuntary muscle tissue)

Mutation - Any alteration in a gene from its natural state; may be disease causing or normal variant.

Mute – A dumb and silent person

Myalgic Encephalomyelitis (*ME*) - Is a severe, complex neurological disease that affects all body systems; also referred to in the literature as chronic fatigue syndrome (CFS), reduce the ability to exercise after suffering from a viral infection

Mycobacterium -Both leprosy and tuberculosis, caused by Mycobacterium Leprae and Mycobacterium tuberculosis respectively

Mycotic aneurysm - an infected aneurysm caused by fungi.

Myopia - a common condition in which light entering the eye is focused in front of the retina and distant objects cannot be seen sharply. In high myopia the eyeball is unusually long, whereas in physiological myopia the eyeball length is normal but the power of the cornea is too great for the axial length.

"Not only was the Teacher wise, but he also taught knowledge to the people; he carefully evaluated and arranged many proverbs." (Ecclesiastes 12:9)

NHS (National Health Service) – Government funded health service for the general community.

NHS Direct – Telephone advice lines manned by nurses and doctors to advise the general public, 24hours, 7days a week

Naked – Without covering or disguise

Name – Word(s) by which a person or thing is identified or known

Napkin – Used for wiping the hands; disposable material secured between the legs to absorb urine and possibly faeces

Narrow Minded – Someone that is not sympathetic towards other people's feelings or ideas

Native – Belonging to a place of birth or the origin of a place

Navel – The small soft hollow grove on the external area of the stomach

Neck – That part between the head and body

Needs – The necessity to maintain and live independent lifestyle to maintain certain standard in a life time; to be in want of something or someone

Neglect by others (*Abuse*) – The unintentional or intentional failure to meet the basic needs of an individual

Negligence – To be careless

Negotiate – To bargain for

Neoplasm – (a kind of tumour); any new and abnormal growth, specifically one in which cell multiplication is uncontrollable and progressive

Nerves – Bundle of fibres that carry messages between the brain and other parts of the body

Nervous – To feel agitated and anxious

Nervous System – Network of nerve cells known as neurones; it carries messages to the brain through the spinal cord to parts of the body

Neurology – This is the studying of the nervous systems

Nice – To be pleasant

Nick – The slang to steal

Nipple – That part of the breast though which a baby feeds from the mother by sucking

Nit - The egg or young of a parasitic insect, such as a louse; the insect itself when young

Nitty-gritty – The basic details

Nod – A quick shake of the head in forward movement

Noise – A loud and disturbing sound

Nominate – To select; appoint

Nominee – Someone appointed for a particular task(s)

Non-Verbal – Communicating without using words

Norovirus - Is a very contagious virus, causing infection from person to person. The infection comes from contaminated food, water and by touching contaminated surfaces.

The virus causes stomach or intestines or both to get inflamed (acute gastroenteritis). This leads to stomach pain, nausea, and diarrhoea and to throw up.

Anyone can be infected with Norovirus and get sick as many times in a lifetime. The illness can be serious, especially for young children and older adults.

Norovirus is the most common cause of acute gastroenteritis in the world.

Norm – A general acceptable level of behaviour pattern

Nose – The front part of the head with two holes where we breathe in and exhale air to the lungs; the nose of a male is larger than that of a female

The nose has specialised cells which are responsible for smelling, conditioning of inhaled air, warming and making it more humid. The hairs inside the nose prevent large particles from entering the lungs

Nonentity – Someone of no importance

Notice board –A board place in a significant public space where written announcements are place for all to see

Nuisance – An annoying person or thing; offensive

Numb – Cannot feel or move

Nurse – Professionally qualified person trained to care for people in different settings

Nutmeg – Hard seed originator of an East India tree; spice used in food

Nutrient - is a chemical needed by an organism to live and grow; a substance used in an organism's metabolism which has to be taken in from its environment

Nutriment – Nourished food

Nylon – A material used for making clothes, stockings and ropes

Nymphomania – A mental illness; an abnormally excessive and uncontrollable sexual desire in women

"**O**f what use is money in the hand of a fool, since he has no intention of acquiring wisdom?" (Proverbs 17:16)

Oats – (Avena sativa); a type of cereal grain People use them as food for themselves and animals e.g. chickens and horses.

Obese – A person is considered an obese when their weight is 20% or more above normal

Obey – To comply with orders

Obituary – Written article that reports recent death of a person, giving an account of the person's life and information about the upcoming funeral.

Objectivity – Ability to view and describe something without being influenced by own feelings and prejudices

Observe – To take note of (as awareness)

Obsess – desires of (a person); to haunt in mind, esp. to an abnormal degree

Obstinate – not yielding to argument or persuasion; difficult to subdue

Obstruct – To stand in the way

Obscure – hidden by darkness

Occupational Therapist - Treat patients with injuries, illnesses and disabilities through the therapeutic use of everyday activities. They help these patients to recover and maintain their daily living and working skills

Occupy – To take up a place

Oculo-cerebro-renal (OCRL) syndrome-(*Lowe Syndrome (LS)* – **Oculo-cerebro-renal syndrome** (also called **Lowe syndrome**) is a rare recessive disorder characterized by congenital cataracts, mental retardation, aminoaciduria and low-molecular-weight proteinuria. Glaucoma is known to be present in about 50% of these cases

Odour – The property of a substance that gives it a characteristic scent

Oesophagus – a muscular tube through which food passes from the pharynx to the stomach

Oestrogens or **estrogens** – (Greek *οἶστρος* (oistros), meaning "gadfly" - The general name for a group of hormone compounds; a general term for female steroid sex hormones that are secreted by the ovary and responsible for typical female sexual characteristics sexual passion or desire) group of compounds named for their importance in both menstrual and estrous reproductive cycles; the primary female sex hormones. It is the main sex hormone in women and is essential to the menstrual cycle; estrogen may also be used in males for treatment of prostate cancer.

Of One's Mind – In agreement

Off – From a place; away for a position

Off colour – Not in good health

Offend – To annoy or anger; causing resentful displeasure in someone

Offer – Put forward for acceptance or refusal

Office – Building where business is conducted by a professional person

Off-licence – A store that sells alcoholic beverages for consumption off the premises

Old age – The later stage of life

Older People – Mixed gender of the human race aged 60years plus

Omelette – A dish made from beaten eggs, cooked with oil in a frying pan, sometimes folded around a filling such as cheese or vegetable

Omission - Something left out, not done, or neglected: an important oversight

On One's mind – Troubling someone

Open – Not closed, as in a doorway by a door and a window by a sash

Open questions – Questions that are worded in a way that invites a full response in answer

Ophthalmic Optician – A health care professional concerned with the health of the eyes and related structures, as well as vision and visual systems

Opinion – A personal view or belief

Opportunity – Chance of a life time; a good chance as to advance oneself

Oppose – To be against someone; to resist

Oppress – To rule harshly (tyranny)

Option –The right to choose

Optional - Not compulsory; possible

Optimised Interaction – Ways of ensuring that Client and Carer relationship is improved. It optimises effective interpersonal interaction and evaluation with regards to Posture and Positioning

V. K. Leigh

(*sitting and standing*), Gestures (*nodding, hand movements*), Eye contacts (*looks*), Expression of the Face (*smiling*); an observed way of confidence building

Organ - collection of tissues joined in a structural unit to serve a common function; an independent body part that performs a special function; though working as a whole, each part has its own function e.g. the heart which is made up of muscle called cardiac muscle

Organic brain syndrome (*OBS*) - Is generally used to describe a decreasing mental function due to medical disease; a psychiatric illness, often, but incorrectly used with dementia

Organisation – A body of people working together for a purpose

Orthopaedic –This is the branch of surgery concerned with conditions involving the proper diagnosis and treatment of injuries and diseases of the musculoskeletal system

Osmosis – The process by which movement of a solvent (as water) passes through a semi permeable membrane (as of a living cell) into a solution of higher solute concentration

Osteoarthritis (*degenerative joint disease*) - the wear-and-tear form of arthritis; joint pain that comes with wear and tear. It can occur in almost any joint in the body but more common in the joints of the hips, knees and spine; affects the fingers, thumb, neck and large toe of the body.

Osteopenia - is a condition in which bone mineral density is lower than normal. It is considered by many doctors to be a forerunner to osteoporosis.

Osteoporosis - (*"porous bones", from Greek: ostoun meaning "bone" and poros meaning "pore"*) is a disease of bones in which bones become fragile and more likely to fracture; Lack of calcium and vitamin D; it is a disease of the bones that happens when a person loses too much bone, makes too little bone or both. As a result, the bones in the body become weak and may break from

116

a minor fall, in serious cases, even from sneezing or bumping into furniture; a condition of fragile bone with an increased susceptibility to fracture. It weakens bone and increases risk of bone's breaking.

Ostracise –to exclude or banish (a person) from a particular group, society; to be excluded through social rejection.

Other – The remaining one of two or more: the other eye

Out – Not in; away from

Outbreak – A term used to describe an occurrence of possibly a war or disease greater than would otherwise be expected at a particular time and place

Outbuilding – A building separate from but accessory to a main house e.g. stable

Out of Mind – Not in someone's thoughts

Out of One's Mind – Insane

Outfit – A set of clothing; set of equipment especially for the practice of a trade

Outgoing – Going away; relinquishing a position, or office e.g. the outgoing chairperson.

Outline – Something that marks the outer limits of an object or figure: a boundary

Outlive – To live longer than everyone in the family; to survive

Overdose – More than the recommended amount of something e.g. a drug. An overdose may result in death or may cause harm

Overdraft – Excessive withdrawal from an account, resulting in a minus balance

Overcoat – The outermost garment usually extends below the knee, but are sometimes mistakenly referred to as Topcoat

Overdue – Beyond the anticipated time; it should have come about sooner: overdue

Overhear – To hear someone talking without their knowledge

Overnight – Remaining during a night; staying the night due to bad weather

Overweight – Excessive weight; too heavy

Owe – In debt to; to have a moral obligation to render

Overspend – Spending more than one can afford

Oxygen - A colourless, odourless, gaseous element constituting about one-fifth of the volume of the atmosphere and present in a combined state in nature. It is a chemical element with symbol O and atomic number 8.

"Plans fail when there is no counsel, but with abundant advisers they are established." (Proverbs 15:22)

P.A.T test – (Portable Appliance Testing), Safety tests carried out on electrical equipments, especially in residential care homes, hospitals, sheltered housing, etc.

P.E.C.S – (Picture Exchange Communication System) – Mainly used for someone with autism that may have little or no ability to use sign language or talk; Pictures are used or shown in exchange of what is needed instead

Pad – Firm but soft cushion-like material used for stuffing; used for containing urine or faeces on incontinent patients in hospitals and care homes

Pain – There are two types of pain namely: -

> *Chronic, persistent or long-term pain* that could last for months or longer
> *Acute pain (short-term pain)* – new pains or recent pain just starting

Pairing-up – Two people working together in providing personal care to a vulnerable or sick person

Pale – To lose colour

Palm – The inside surface of the hand

Palpitation – Rapid beating of the heart due to an illness or disease

Palsy – Paralysis

Panchromatic – Someone that is sensitive to light of all colours

Pancreas – Large gland situated under and behind the stomach, secretes digestive fluids to aid food to digest in the intestines

Pandemic – An epidemic of disease spread over a large geographical area and affecting the mass population

Panegyric - A formal public speech, or written verse, delivered in high praise of a person or thing, a generally highly studied and discriminating eulogy, not expected to be critical.

Panel – List of people or jury chosen for a purpose

Panic – This is fear that spreads from one person to another

Paramedic – Someone trained in emergency and possibly medical procedures

Parasympathetic Nervous System - Part of the involuntary nervous system that slows the heart rate, increases the intestinal and glandular activity, and relaxes the sphincter muscles. The parasympathetic nervous system, together with the sympathetic nervous system, constitutes the autonomic.

Parkinson's disease - *(PD also known as idiopathic, hypo kinetic rigid syndrome)* - is a degenerative disorder of the brain that leads to shaking (tremors) and difficulty with walking, movement, and coordination. It can be contained for a short period of time but cannot be cured; another disease related to the nerve cells, and its causes are yet unknown. Dementia affects the mental health of the individual whereas Parkinson's disease is primarily a physical disability. The symptoms of Parkinson's are uncontrollable shaking of the limbs, that is just as frustrating. There is no cure for this disease, although it may be contained to some extent for a short period, with the use of drugs. After a certain period, the symptoms cannot be controlled.

Partnership – Working together as equals

Partnership approach to assessment – The way of working that makes an individual the centre of the process; it is important to have partners in care because – professionals have the knowledge, there is shared responsibilities, carers know the available resources

Pathogen - is a microorganism —in the widest sense, such as a virus, bacterium, fungus —that causes disease in its host which may be an animal (including humans), a plant, or even another microorganism.

Patience – Ability to suffer annoyance; the bearing of provocation, annoyance, misfortune, or pain, without complaint, loss of temper, irritation

Pre-osteoporosis (*Osteopenia*) - The loss of some bone density; if allowed to proceed, the condition could progress to being osteoporosis

Primary Care Act (1997) – The ways in which primary care is delivered (NHS) e.g. dental, medical, ophthalmic services; contracts with NHS and medical practices

Pedicure – Treatment of Bunions, corns, toenails by a chiropodist

Pendant – An ornament on a chain; used by the vulnerable to call for assistance in hospitals and care homes

Penicillin – Substance extracted from moulds used in medicine to stop the growth of many bacteria

Penile cancer - malignant growth found on the skin or in the tissues of the penis.

Penis – The part of the male body used for urinating

Pension – Money given regularly by the state for services rendered or due to old age

Pericarditis - an inflammation of the lining that surrounds the heart; a rare condition usually caused by an infection.

Peri-menopause - The time around menopause. It refers to the menopausal transitional period. Peri-menopause is **not** officially a medical term, *but* is sometimes used to explain certain aspects of the menopause transition in lay terms.

Periodontitis - is inflammation and infection of the ligaments and bones that support the teeth.

Permit – To give permission; to allow

Pernicious anaemia -Is a decrease in red blood cells that occurs when your intestines cannot properly absorb vitamin B12.

Persistent diarrhoea - often causes nutritional problems, creating the risk of malnutrition and serious non-intestinal infection. Dehydration also occurs; the main danger is malnutrition and serious non-intestinal infection; dehydration may also occur

Persistent Sexual Arousal Disorder - *Persistent genital arousal disorder (PGAD)*, originally called *persistent sexual arousal syndrome (PSAS)* and known as *restless genital syndrome (ReGS or RGS)*, results in a spontaneous, persistent, and uncontrollable genital arousal in women, with or without orgasm or genital engorgement, unrelated to any feelings of sexual desire; the syndrome of priapism in men to be the same disorder.

Masturbation and orgasms offer little or no relief. It is a very rare condition, which sufferers don't report the condition because of the shame and embarrassment the disclosure would involve.

Person – Someone outer part of the body

Personal Protective Equipment (*PPE*) – Items commonly used in health and social care settings e.g. Plastic gloves, paper masks, paper hair cover, plastic aprons, paper gowns, plastic goggles and plastic overshoes

Personal development - Activities that improve awareness and identity; develop talents and potential; build human capital and facilitate employability enhance quality of life and contribute to the realization of dreams and aspirations.

Personal development plan – *IT IS PERSONAL* – The structure that helps someone to reflect and appraise their working and learning performances, planning on how to develop in stages towards future goals

Personalisation – Often associated with direct payments and personal budgets, under which service users can choose the services that they receive, personalisation also entails that services are tailored to the needs of every individual, rather than delivered in a "one-size-fits-all" fashion for every person who receives support, whether provided by statutory services or funded by themselves, will have choice and control over the shape of that support in all care settings".

Personal Protective Equipment (*PPE*) – Protective equipments, clothing, disposable aprons and gloves used to keep away bacteria and to create barrier against pathogens that cause disease; PPE is a very important ways of infection prevention and control in the health and social care settings

Employer Responsibilities	Employee Responsibilities
Must Supply the correct PPE	Must use PPE at the right times
Must maintain PPE & store correctly	Must report PPE if faulty or low in stock
Must make available –instructions, training & notices on PPE	Must attend training and follow instructions on PPE
Must carry out risks assessments to inform decisions on PPE	Must carry out and follow risk assessments on the right PPE for task
Must ensure that PPE is being used properly	Must use PPE whenever the need should arise and NEVER use a short cut
	Must dispose of PPE in the right manner

The Responsibilities and Roles of Employer and Employee on Personal Protective Equipment (PPE)

Personality – The characteristic of a person

Persuade – To convince someone

Pertussis (*whooping cough*) — is a highly contagious bacterial disease caused by *Bordetella pertussis*. It is called the cough of 100 days.

The bacteria Bordetella pertussis

Its symptoms are initially mild then develop into severe coughing fits, which produce the namesake high-pitched "whoop" sound in infected babies and children, when they inhale air after coughing. It lasts for approximately six weeks before subsiding.

It can be prevented by vaccination; antibiotics are recommended because they shorten the duration of infection. It is known that the disease affects about 48.5 million people worldwide and kills about 295,000 people yearly.

Pervert – Abnormal

Petition – A written and formal request

Pharmacy – A shop where medicines are prepared

Petty-cash – Little amount of money set aside for minor expenses

Pharynx – The tube connecting the mouth with the tube to the stomach

Philtrum - The vertical groove on the median line of the upper lip on the human face

Phimosis - Is defined as the inability of the prepuce (foreskin) to be retracted behind the glans (*vascular structure located at the tip of the penis in male mammals*) penis in uncircumcised males.

Phlebotomists –These are people trained to draw blood from a patient for clinical or medical testing, transfusions, donations, or research

Phlebotomy (Greek word *phlebo-*, meaning "pertaining to a blood vessel", and *-tomy*, meaning "to make an incision") is the process of making an incision in a vein with a needle. The procedure itself is known as a venipuncture.

Phobia – Fear of something without reason

Physical (*Abuse*) – The use of force, resulting in any type of injury (pain, cuts, impairment, etc.) to the body

Pill – A tiny lump of medicine that must be endured to heal or relive medical condition

Pinafore – A type of apron worn to cover the front of a dress

Pinch – To hurt by squeezing the flesh or any object between the thumb and finger

Pitch of voice – The is the conversational tone of voice of an individual; it can be too high or too low - each person's vocal pitch is different

Plague – Deadly disease carried by rat flea; nuisance

Pneumonia -A common lung infection caused by bacteria, a virus or fungi. Pneumonia and its symptoms can vary from mild to severe; coughing, fever, shortness of breath, and chest pain. The second stage before Sepsis, The final stage of high fever.

Podagra – Known as "gout" or uric acid arthropathy, is a rheumatic complaint, which usually attacks a single joint at a time.

The disease also called as the "disease of the rich" or the "king's disease, is a disease that occurs due to the elevated levels of uric acid in the blood causing joint inflammation.

Police – Men and women employed to enforce the law

Policy – A course of action decided on; a rule to guide decisions and achieve rational outcomes. A policy is a written statement of intent, and is implemented as a procedure or protocol

Polite – Respectful; showing good manners toward others e.g. in behaviour, speech, etc.

Polygamy – Having more than one wife or husband at the same time

Polymer - a large molecule composed of many repeated subunits, known as monomers. Because of their broad range of properties both synthetic and natural polymers play an essential and ubiquitous role in everyday life.

Polymyalgia Rheumatica (*PMR*) – A rheumatic condition in which people have many (*poly*) painful muscles (*myalgia*). Cause is unknown and usually treated with drugs called steroids (*corticosteroids*). PMR can start from the age of 50years onwards but more from age 70years and affects women 2 – 3 times as often as men.

Polysaccharides – These are polymeric carbohydrate molecules composed of long chains of monosaccharide units bound together by glycoside bonds.

Poppycock – Nonsense

Porridge – Food made form oat (wheat) with milk and water

Positive discrimination – Discrimination favourable to those that are at disadvantage

Postmenopausal - is used to refer to the time *after menopause* has occurred, e.g. doctors may speak of a condition that occurs in "postmenopausal women." These are women who have already reached menopause.

Post- mortem – Examination of a dead body to establish the cause of death

Post-Traumatic Stress Disorder-*PTSD* is an anxiety disorder that some people get after seeing or living through a dangerous event. Anyone can get PTSD at any age. This includes war veterans and survivors of physical and sexual assault, abuse, accidents, disasters, and many other serious events. It is a chronic condition that never goes away

Power – The capacity, strength and ability to having control and influence over another person

Practice - Customary performance; making a habit of...

Practitioner - Someone who practices an occupation or profession e.g. Doctor, Engineer, Law, etc.

Presentation – The act of introducing by speech and various additional means e.g. by electronic means, sharing computer screen or projecting some screen information

Pressure group - Are individuals who have same ideas and beliefs based on ethnicity or a common goal; they take action based on their beliefs to promote change and further their goals. They represent viewpoints of people who are dissatisfied with the current conditions in society and alternative viewpoints that are not well represented in the mainstream population

Prejudice – An unreasonable or unfair hatred or preference towards a person or a group of people

Prepuce - homologous structures of male and female genitals; the fold of skin that covers the head of the penis; foreskin; similar covering of the clitoris.

Priapism –This is a potentially painful medical condition, in which the erect penis does not return to its flaccid state, despite the absence of both physical and psychological stimulation, within four hours.

V. K. Leigh

There are two types of *Priapism*: **Low-flow** and **High-flow**;

Low-flow -involves the blood not adequately returning to the body from the organ. 80% to 90% of clinically presented priapism is low flow disorders.

High-flow- involves short-circuit of the vascular system partway along the organ. Treatment is different for each type.

A medical emergency, which should receive proper treatment by a qualified medical practitioner; early treatment can be beneficial for a functional recovery, an overdose of the drug Sildenafil or Viagra that cause pro-long erection in a man: it must be treated within 24hours else it may cause permanent damage to the male reproductive organs.

Priapism may be associated with blood disorders, especially sickle-cell disease, sickle-cell trait, and other conditions such as leukaemia (cancer of the blood or bone marrow) - a treatable disease

Priapism may also be associated with abnormally low levels of glucose-6-phosphate dehydrogenize (G6PD)

Priapism is a persistent penile erection without sexual desire. Put simply, the male has an erection that does not go away. Treatments for priapism include drugs, drainage of blood from the penis, or anesthesia.

Priapulid – *Priapulid***:** is a phylum of marine worms; an invertebrate phylum of wormlike marine animals; the body is made up of three distinct portions (proboscis, trunk, and caudal appendage)

(4 to 6 inches) and is often covered with spines and tubercles, and the mouth is surrounded by concentric rows of teeth. There are 17 species of predatory, unsegmented marine worms that live in the sand and mud at the sea bottom.

Prions - A disease representing conditions that affect the nervous system in humans and animals.

In humans, these conditions impair brain function, causing changes in memory, personality, and behaviour; a decline in intellectual function known as dementia, abnormal movements, difficulties with coordinating movements (ataxia). a very rare. Disorder where the affected person inherits the affected gene from one parent.

Principle – Rule; an accepted rule of action or conduct: a person of good moral principles

Privacy – Secrecy; personal

Privacy Law - General privacy laws have an overall bearing on the personal information of individuals;

The Privacy Act of 1974 requires that agencies create and maintain, as necessary, System of Records Notices (SORN) as defined in the Privacy Act. A system of records consists of any item, collection, or grouping of information about an individual, where those records can be retrieved by the name of the individual or by some other type of identifier unique to the individual.

From - (www.hhs.gov/foia/privacy/index.html)

Procedure – The documents that show ways of dealing with specific issues; set of commands that show how to prepare or make something

Professional – Expert; a person who is engaged in a certain activity, or occupation, for gain or compensation as means of livelihood

Progeria - (*Hutchinson–Gilford Progeria syndrome, HGPS, Progeria syndrome*) An extremely rare genetic disorder wherein symptoms resembling aspects of aging are manifested at a very early age; a disorder that causes sufferers to age about eight times faster than their natural ageing process,

Prohibition – Prevention; to stop something from happening

Project – A task, piece of research; scheme that requires a large amount of time, effort, and planning to complete

Prolactin - A peptide hormone (*PRL* gene), known as lacto-trope, a protein in human that is best known for enabling female mammals to produce milk. It is influential over a large number of functions with over 300 separate actions of PRL, reported in vertebrates; peptide hormones released by the pituitary gland from the anterior, or adenohypophysis lacto- trope cells, stimulating mammary glands.

Prostate (*Greek - προστάτης, prostates, literally means, "one who stands before", "protector", "guardian"*) - A compound tubuloalveolar exocrine gland of the male reproductive system in most mammals; a healthy human prostate is said to be slightly larger than a walnut while in the female body, the paraurethral glands, or Skene's glands are the female prostate.

Prostate Cancer - A form of cancer that develops in the prostate, a gland in the male reproductive system; it is usually a slowly-progressing disease and only affects men

Prostatitis – This is a painful infection of the prostate gland. It causes fever, chills, painful urination, lower back pain, pain in the genital area and there are different types namely acute bacterial Prostatitis, chronic bacterial Prostatitis, chronic pelvic pain syndrome and Asymptomatic Inflammatory Prostatitis

Principles of good practice – *Care* – Principles agreed for workers in health and social care; such as promoting values of equality and diversity, developing and maintaining relationships with clients but maintaining the boundaries, maintaining information confidentiality provided by clients, allows clients to be heard, freedom of speech, etc.

Psoriasis – A common skin disorder that causes skin redness and irritation. Most people with psoriasis have thick, red skin with flaky; it is typically a lifelong condition. There is currently no cure – research for cure is still in progress at time of publication

Psychiatric – The area in medicine, dealing with the study and treatment of mental illness

Psychopath – A person who is not morally responsible

Psychological /Emotional (*Abuse*) – Actions taken by another person that damages someone's mental wellbeing

Psychosis – An illness of the mind

Psychotic – Someone suffering from psychosis

Psychotic disorder – A serious mental illness that make sufferers to lose contact with reality

Public Information Signs (P.I.S) – This is used for people who do not speak English with difficulties in reading and writing

Pull-cord – Draw strings used in flats for the vulnerable to call in times of emergencies, commonly found in hospitals, residential care homes, sheltered accommodations and doctors' surgeries.

Pulmonary hypertension (PH) - is an increase of blood pressure in the lung vasculature (***pulmonary artery, pulmonary vein, or pulmonary capillaries***), leading to shortness of breath, dizziness, fainting, leg swelling and other symptoms; a severe disease with a marked decreased exercise tolerance and heart failure.

Pulse - Someone's **pulse** represents the palpation of the heartbeat using the thumb and fingertips; it may be palpated in any place that allows an artery to be compressed against a bone, e.g. the neck, the inside of the elbow or at the wrist

Purge – To empty the bowel

Pus – Thick yellowish fluid from wounds

Putrid – Rotten; very bad

Pyloric sphincter - Located at the base of the stomach and is the contracting ring of muscle which guards the entrance to the small intestines. It lets food pass from the stomach to the duodenum.

"Quick-tempered person stirs up dissension, but one who is slow to anger calms a quarrel." (Proverbs 15:18)

Quack – Someone who pretends to have the qualification and skills that they do not have; a "fraudulent pretender to medical profession

Quadruplet – Four children born at the same time by one mother; a multiple birth when more than one foetus is carried to term in a single pregnancy.

Quality - A measure of excellence; something that can be noticed as a part of a person or thing

Quarantine – Compulsory confinement to prevent e.g. the spread of disease; strict isolation imposed to prevent the spread of disease.

Quarrel – Angry dispute; disagreement within

Queen – A female monarch; wife of a king

Queer – Odd person; a term for sexual and gender minorities that are not heterosexual or gender-binary

Quench – To satisfy one's thirst; to put out fire with water

Query – Doubt; a question in the mind

Question – The procedure of asking; used in many aspects of Care (when preparing care plans)

Questionnaire – Questions used to collect information; a good way to gather information from a group of people.

Queue – People in line waiting for their turn

Quick-tempered – Someone who gets angry quickly; easily upset

Quid-pro-quo – (***Latin***) one thing in return for another; something given as an exchange

Quinine – A bitter substance (liquid) used for the treatment of malaria

Quintuplet – Five children born by one mother at the same time

Quit – To leave (as in a job)

Quiver – To tremble over someone or thing

Quorn - A vegetable substitute for meat; makes meat substitutes that actually taste like meat!

Quota - Share which each one is called upon to contribute; a division to each member of a body (an allotment)

Quotation – A pricing estimate; something that is acknowledged; a passage quoted from a book or speech.

"**R**eceive my instruction rather than silver, and knowledge rather than choice gold." (Proverbs 8:10)

RNA virus - a virus that has RNA (ribonucleic acid) - is a virus as its genetic material.

This nucleic acid is usually single-stranded RNA (ssRNA), but may be double-stranded RNA (dsRNA).

Notable human diseases caused by RNA viruses include **Ebola hemorrhoragic fever**, SARS, influenza, hepatitis C, West Nile fever, polio, and measles.

Racemose Aneurysm - dilatation and tortuous lengthening of the blood vessels.

Race Relation Act (1976) – This is an Act of Parliament condemning general discrimination against anyone on grounds of colour, nationality, religion, race or ethnic group; *discrimination against employment (Industrial tribunal); on racial grounds (County court)*

Race Relations (Amendment) Act 2000 – This amendment came into effect following the Stephen Lawrence case in which the police work was affected by *Institutionalised racism*

Radiator – Equipment used for heating a room to a required temperature

Rag – A piece of cloth usually used for cleaning

Rage – To be very angry; violet anger

Rampage – Someone rushing about with anger inside of them

Rape – Forceful act of sexual intercourse with someone against their consent

Rapport - A relationship of mutual understanding or trust and agreement between people

Rat – A destructive animal like a mouse but larger

Rate – The standard of payment like in wages where some staffs are paid at higher rates compared to others; an annual tax paid by property owners to the local council

Reaction – A process that leads to the transformation of one set of chemical substances to another in animals and plants

Readable – Clear enough to be read

Records – Documented information of individual client for future use (access to these files are strictly confidential)

Recover – To find again (to recover a missing tea cup)

Reference – A note from a person, confirming details of the individual

Reflective practice - The capacity to reflect on action in order to engage in a process of continuous learning

Reflect - Think deeply about an issue.

Reflecting-in-action – This takes place when an issue arises or when an event takes place

Reflecting-on-action – This is when consideration is being given after a job had been well done and how the situation would have been managed in a different way

Reflective practice – To be mindful, give oneself time and space to carefully think about how someone works

Reflective Practice Process – The processes that saves time and when it is used productively – *see diagram for details'*

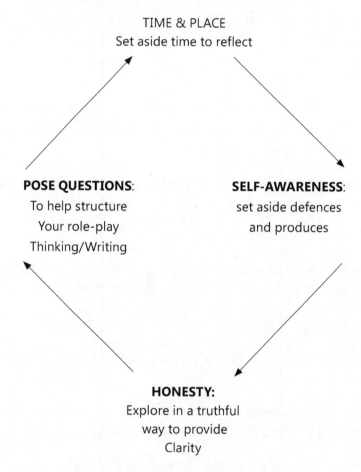

TIME & PLACE
Set aside time to reflect

POSE QUESTIONS:
To help structure
Your role-play
Thinking/Writing

SELF-AWARENESS:
set aside defences
and produces

HONESTY:
Explore in a truthful
way to provide
Clarity

The reflective practice process

Register – A written record

Regulations – Set of laws; law or order prescribed by authority, especially to regulate conduct; using the authority of a statute

Rheumatoid Arthritis - (**RA**) is an autoimmune disease that results in chronic systemic inflammatory disorder that may affect many tissues and organs, but principally attacks flexible (synovial) joints. It can be a disabling and painful condition, which can lead to substantial loss of functioning and mobility if not adequately treated.

Relevant - Having a bearing on the matter at hand; upon or connected with the matter in hand

Relevant Standard – The standards for helping the healthcare industry improve efficiency and quality for lower costs; more streamlined care delivery and better patient safety.

Reporting of Injuries, Diseases and Dangerous Occurrences Regulations (*RIDDOR*) 1995 -Employers to make known to the health and safety executive, range of occupational injuries, diseases and dangerous events

Research –To undertake creative work on a systematic basis in order to increase the stock of knowledge

Respect - *High opinion* - expressions of high or special regard

Respiratory system - The biological system that introduces respiratory gases to the interior and performs gas exchange; to supply the blood with oxygen in order for the blood to deliver oxygen to all parts of the body

Respond - To reply or answer in words: to a question.

Responsibility – The actually expectation required of a person, legally

Restless Legs Syndrome - *(RLS)* - is a disorder of part of the nervous system that causes an urge to move the legs. It usually interferes with sleep and a sleep disorder. Commonly diagnosed in women, can begin at any age and in young children; people affected are of middle-age or older. Also known as **Willis-Ekbom Disease (WED).** Treatment of disorder can cause insomnia.

Resuscitation – Emergency method to restore breathlessness (Lack of breathing) which will involve the *A – Air, B – Breathing and C- Circulation (A, B, C)* of life.

Resourcing of service - The professional business function that runs the overall affairs of an organization's human resources is called human resource management (HRM, or HR)

Retribution – A *"pay-back"* scheme that could cause an injury devised by someone to retaliate for what had been done to the person in the past

Require - To have the need of; to call on authoritatively

Rheumatoid arthritis - Causes pain in the feet, wrists, hands, and other joints; it is a long-term disease that leads to inflammation of the joints and surrounding tissues. It can also affect other organs.

Rights – A person's legal and moral entitlement

Risk – The taking a chance of losing, damaging or experiencing disaster; a client or official can be exposed to harm from the environment that they find themselves; the chance of harm being done by a hazard e.g. wet floor, drawing pin on the floor.

Risk Assessment – The method by which a care setting is examined for potential hazards or dangers to the clients and carers. The areas of concerns are usually addressed and documented for record purpose and measures taken to reduce the risks to a safer level

Risk Assessment process – The steps involved in a risk assessment process in relation to infection control

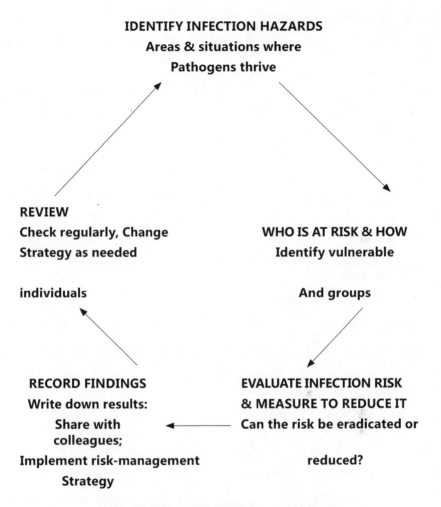

IDENTIFY INFECTION HAZARDS
Areas & situations where
Pathogens thrive

REVIEW
Check regularly, Change
Strategy as needed

individuals

WHO IS AT RISK & HOW
Identify vulnerable

And groups

RECORD FINDINGS
Write down results:
Share with
colleagues;
Implement risk-management
Strategy

EVALUATE INFECTION RISK
& MEASURE TO REDUCE IT
Can the risk be eradicated or

reduced?

THE RISK ASSESSMENT PROCESS

Risk Management – This is when management identifies health and safety hazards' and implement ways to deal with it (identifying and assessing the risks, implementing strategies to control it and providing finance to sort it out)

Regulations for Reporting of Injuries, Diseases and Dangerous Occurrences (RIDDOR) – This is the regulations which relates to the recordings and reporting of ill-health and accidents at work. Serious accidents and related dangerous occurrences must be documented and reported within seven days.

V. K. Leigh

Role – The part played by an individual when involved in social environment e.g. Doctor, social worker, nurse or ambulance personnel

Role play – This is when a scenario is being acted for a better understanding, things that happen between two or more individuals

"Say to wisdom, "You are my sister," and call understanding a close relative" (Proverbs 7:4)

Saccharine – A sweet substance used in place of sugar

Sad –Not happy; **Seasonal affective disorder** (*SAD*), also known as seasonal depression, this is a mood disorder

Safe – To be secured from injury or danger

Safe guard – To protect from danger

Safeguarding – The protection of adults who are more vulnerable to harm and abuse; protection of individuals from illness, injuries and abuse, helping to promote their interests, values and upholding their rights

Safeguarding Vulnerable Groups Act 2006 – The Sir Michael Bichard Inquiry 2002 (*On 4 August 2002, two English girls were murdered in the village of Soham, Cambridgeshire. The victims were Holly Marie Wells and Jessica Aimee Chapman, both aged 10 – the Soham murders-*) The recommendations led to the creation of Independent Safeguarding Authority (ISA) now responsible for – *the Vetting and Barring Scheme (VBS)* and *Maintaining children and adults barring lists*

Safe practice – Ways of working, to uphold laws and standards, to protect, prevent and promote safety and the wellbeing of all others

Sahib – Name given to foreign dignitaries in India as a term of respect

Salad – Cut up mixed vegetables (raw or cooked) for eating, usually cold dish; consisting of vegetables e.g. lettuce, tomatoes, and cucumbers, covered with dressing

Salmonella – These are bacterium, a special type of bacteria that cause food poisoning

Saliva – A digestive fluid from the mouth

Saliva gland – The gland in the body that produces saliva from the mouth

Salute – Greetings of honour; courtesy

Sane – To be mentally sound and able to do things or be involve

Sanitary – As of health, free from elements such as dirt and agents of infection or diseases

Santa Claus – A legendary also known as St Nicholas - father Christmas who brings gifts to children

Sarcasm – Remarks made in order to scorn someone

Sardine – Young Pilchards (***fish***) tinned and flavoured in tomatoes sauce or in oil

Satan – Known as the deceiver; the devil and head of evil and temptation

Saucy - Cheeky

Save - To be taken out of danger

Saver – Someone who is mindful of spending

Scabies – Skin disease caused by mite

Scandal – Damages to someone's reputation

Scar – Mark left behind from an old injury

Scarlet fever – An infectious fever that causes sore throat and rashes on the skin

Scare – Unexpected fright; to startle someone

Scarf – A loose material worn around the neck or shoulders as a part of dressing

Scent – Natural odour from a person; cause to smell pleasant

Sceptic – Someone that always doubts

Scheme Managers – Employed by councils and housing organisations to run and maintain sheltered accommodations

Schizophrenia – Split personality; a form of madness; mental illness that causes delusions, hallucination and physical instability

Sciatic – Region of the hip

Sciatica – A general pain given to any pain that is caused by irritation or compression of the sciatic nerve. This nerve is the longest nerve in the human body; made up of vertebrae, discs and nerves - *vertebrae is made up of a set of bones that make up the structure of the human spinal cord and protecting the nerves* - running from the back of the pelvis, through the human buttocks, travelling along the back areas of both legs ending at the feet: it is a severe pain in the upper part of the leg (thigh) near the hip. Sciatica is different to general back pain; usually causing little problem to the back and could last for weeks or months.

Scratch – A slight wound or injury to the skin

Scrotum – The bag of skin under the male organ containing the testicles

Scrounge – To borrow with no intention of giving back; scavenging

Sebaceous cyst - is a closed sac under the skin filled with a cheese-like or oily material (*see*

Infected sebaceous cyst behind the ear

Seborrheic Keratoses - These are brown or black growths usually found on the chest and back, as well as on the head. They originate from cells called keratinocytes. As they develop, seborrheic keratoses take on a warty appearance. They do not normally lead to skin cancer; Seborrheic keratoses are harmless and are not contagious.

Secretagogue - A substance that causes another substance to be secreted, e.g. Gastrin

Sedate – (Tranquillize) - To dose with sedatives.

Sedative – (Tranquilliser) - A medicine that makes you sleepy

Senility - The state of being senile, especially the weakness or mental infirmity of old age

The mental and physical deterioration associated with aging.

Sense Organs - There are five different sense organs, namely: *sight, smell & taste, touch, and hearing.* Each of the 5 senses consists of organs with specialized cellular structures that have receptors for specific stimuli

"The Sense of Sight"
The sense of sight is the eye; light enters the eye through the cornea and it is sent to the pupil, the dark centre of the eye.

"The Sense of Smell and Taste"
The tongue and nose contain receptor cells that receive information. The receptor cells in the nose send messages thorough the nerves. The tongue can only recognize four different kinds of tastes; they are sweet, salty, sour, and bitter.

"The Sense of Touch"
The skin is the sense organ for touch and because it forms the body, it makes it to be the largest organ with Sense receptors all over the skin receiving different sensations.

The Sense of Hearing

The ear is the sense of hearing; the eardrum vibrations cause three small bones in the middle ear to vibrate. These bones are the hammer, anvil, and stirrup-pass the vibrations to a snailed organ in the inner ear.

Sensitivity - The ability of an organism or organ to respond to; the strength of physical or emotional reaction insensitive people

Sensory loss – A pattern of neurological damage caused by a lesion to a single tract in the spinal cord

Sepsis – (*septicaemia*) This is a blood infection caused by the immune system's response to serious bacterial infection, viruses, and parasites in the blood, tract, lungs, skin, or other tissues; can lead to organ failure and death.

Service - An act of helpful activity; aid

Search – To carefully look over in order to find something

Seat – An object on which one rests the body e.g. chair

Seclude – To be kept apart from the others; to remove

Secret – Something not known about because it was not told

Secrete – To produce; to give off e.g. sweat gland produce sweat

Section – To separate; remove by division; admitted by force into a hospital as a result of mental health

Secure – Made safe; free from danger

Self- confidence – Trust in oneself; believe in you

Self-directed support – (Personalisation), Individuals have control over their own care and how it is provided, known as an individual budget or direct payment

Self- esteem – Self-respect; self confidence in knowing who someone is

Self-image – The way we see ourselves and our intelligence

Self- referral – a choice by the patient of medical specialists without the need for primary care physician

Seminiferous tubules - Located in the testes, and are the specific location of meiosis, and the subsequent creation of gametes, namely spermatozoa.

Sense organ – There are five sensory organs in the human body, namely vision, hearing, smell, taste, and touch.

Sensory Impairment Team – A team of specialists in various kinds of sensory impairment

Sensory impairments – A disorder which affects the nervous system and sense organs; because of its effect on the nervous system, the body automatically responds thereby affecting the five sense organs of vision, hearing, smell, taste, and touch

Sepsis –A condition of the blood poison caused by bacteria (**wound)**

Service users – Clients using the health and social care services at their disposal

Service requirements – The required care service from health and social care laid down to meet individual needs of a client

Service agreement – The written statement of responsibility between care groups and the NHS trusts which reflects the national standards and targets

Settee – An extended seat with back to rest on

Sew –Using thread and needle to put together

Sexual (*abuse*) – This is the involvement of direct or indirect sexual activity without the person's consent

Sex Discrimination Act (1997) – The Parliamentary Act which makes it illegal to discriminate on the grounds of gender or status in all areas of public offices, e.g. Police (Height recruitment for men and women), ministers of religion, etc.

Sexual intercourse – (coitus; copulation; mating) the process by which spermatozoa from a male are deposited in the body of a female during sexual reproduction; heterosexual intercourse, involving penetration of the vagina by the penis

Sexually Transmitted Diseases – (*STD*), also referred to as *sexually transmitted infections (STI)* and *venereal diseases (VD),* are illnesses that have a significant effect on the body in both women and men; Diseases passed from one person to another

Shampoo – Chemical used for washing the head and hair

Shave – The process by which hair is removed from the body; using a razor

Shawl – A simple item of clothing, loosely worn over the shoulders, upper body and arms, and sometimes also over the head, usually by females in Asia

Share - A part or portion belonging to distributed to or owed by a person or group. An equitable portion: to do one's share of the work; to distribute among each other

Shelf – A flat, usually of rectangular structure, composed of a rigid material, such as wood, glass, or metal, fixed at right angles

Sheltered Housing – A place that gives the feelings of home with some support and comfort to the vulnerable with same age group

Shin – The front part of the leg from the knee to the ankle

Shinbone – (tibia); the larger of the two bones in the lower leg. The shin bone is known as the tibia. Its smaller companion is the fibula.

Shingles (*Zoster*) - the reactivation of the chickenpox virus, dormant since a childhood infection

Shock – Unexpected bad news

Shoulder – the top of the upper arm; the part of the human body between the neck and upper arm

Shroud – Cloth used to cover a dead body

Shuffle –To move without lifting the feet

Sibling – A brother or sister in a family

Sick –To vomit; to joke about things that are very sensitive

Sickle- cell disorders –This is inherited blood disorder commonly affecting African and Caribbean descendants

Signs – A clear indication, seen on someone's appearance that he is not feeling well

Signs and Symptoms - Signs are different from symptoms; *signs* - any indication of a medical condition that can be objectively observed (i.e., by someone other than the patient), whereas a *symptom* is merely any manifestation of a condition that is apparent to the patient (i.e., something consciously affecting the patient).

Sign Language (BSL) – British Sign Language – Used for helping people who are deaf to communicate with others; a method of communication using the hands to convey the message (the deaf)

Sincere – An honest and truthful person

Singe – Surface burns

Sinus – A hollow cavity in the front part of the head, connected with the nose

Sip – Drinking in small quantity

Skeleton – (Greek *skeletos* "dried body") - The body part which forms the supporting structure of an organism; there are two different skeletal types: the exoskeleton which is the stable outer shell of an organism, and the endoskeleton which forms the support structure inside the body; the bony framework of a mammal;

Skills - The ability from one's knowledge, practice and aptitude to do something well:

Skin – The outer covering of someone; the largest organ of the body. The skin protects us from microbes and the elements, helps to regulate the body temperature and allows the sensations of touch, heat, and cold.

There are three layers to the skin, namely: the outermost layer of skin provides a waterproof barrier and creates our skin tone (Epidermis); beneath the epidermis are the tough connective tissue, hair follicles, and sweat glands (Dermis) and The subcutaneous tissue which is made of fat and connective tissue (Hypodermis)

Skin care - Skin care guide will give the information needed to have beautiful, radiant skin.

Skin Tag - A skin tag is a small flap of tissue that hangs off the skin by a connecting stalk. Skin tags are not dangerous. They are usually found on the neck, chest, back, armpits, under the breasts, or in the groin area. Skin tags appear most often in women, especially with weight gain, and in elderly people. Skin tags usually don't cause any pain. However, when anything, such as clothing or jewelry, rubs them it can become irritated; tags have been seen to grow up to a half-inch long they are typically the size of a grain of rice.

Skip –A large size rubbish container

Skull – The bony framework of the head

Slash –A long cut on the body

Sleep – A state of unconsciousness; a psychological process that allows the body to go into rest within a 24 hours' period

Sleep apnoea – is a chronic condition that is ongoing and disrupts the sleep. It causes breathing to become shallow and turns deep sleep into light sleep. The quality of sleep becomes poor, causing tiredness during the day due to excessive daytime sleepiness and gives a poor quality of sleep. This is a common disorder.

Sleeping Sickness – (SS) - *(trypanosomiasis)* - is an infection caused by certain flies called Tsetse. It causes swellings in the brain when an infected fly bites someone; the infection spreads through the blood and it is commonly found in Africa. Its complications include: Gradual damage to the nervous system; Uncontrollable sleep as the disease gets worse. If left untreated, the infection progresses to death within months or years.

Small Intestine - The *duodenum* is the first part of small intestine; the small intestine is where most of the absorption of digested food occurs. Simple sugars and amino acids are absorbed through intestinal epithelium and enter the blood stream to be used by cells

Small talk –Gentle conversation; light discussion

Smoking – A habit whereby someone inhales and exhale the smoke of burning tobacco containing carbon monoxide, tar and nicotine that are harmful to the human health

Smooth muscles – A type of muscles found in the genital, urinary and respiratory tracts and the blood vessels; these muscles are controlled by the hormones and nervous system

Snatch – To quickly pull away from

Sneeze – To blow out air violently and involuntarily through the nostrils of the nose

Snore – Rough and hoarse breathing while sleeping

Snort – To noisily force air through the nostrils; disapproval noise

Social behaviour – The ways people talk and relate to one another in the community

Sociable - Friendly

Social care – Cares being provided in Residential and Day care environment

Social inclusion – The process where all members of the public have access to all available services and activities; it ensures that those prevented through circumstances, get the opportunities that allow them to make decisions and participate in society: thereby improving their wellbeing

Socially excluded - A lack of belonging, acceptance and recognition; socially excluded persons are more socially vulnerable, and hence they tend to have diminished life experiences

Social model – The approach which views mental health as consequences of higher social problems being caused by social factors such as poverty, discrimination and inequality within its society

Social model of disability – When the factors (decision makers), within society determines who is disable and who is not disable

Socialisation – The ways by which every individual learn culture and ways of life

Social Worker – Person responsible to assist in providing welfare services to the vulnerable.

Sodium – A soft compound that forms salt, silvery in colour, also forms compound as washing soda

Sofa – An extended long seat with stuffing to the bottom and back to sit on

Soggy – Damp, wet and heavy

Soil - Dirt

Solicitors – A lawyer who prepares and manages case in the court of law

Solvent – Liquids that dissolve other liquids to form a solution; able to dissolve

Sore – A painful and exposed injury

Speak – To convey thoughts, opinions, or emotions orally

Special Features – Special quality

Speech and Language Therapist *(S.L.T.)* – It is the helping of those with communication problems e.g. Physical disabilities, Severe Learning disability, partial deafness, speech impediments, autism, etc.

Sperm or spermatozoa - The part of the semen that is generative—can cause fertilization of the female ovum.

A mature male germ cell that develops in the seminiferous tubules of the testes; resembling a tadpole and has a head with nucleus, neck and tail that provides propulsion.

Produced in vast numbers after puberty, spermatozoa, the generative component of the semen

Spill – An overflow of liquid to the ground

Splutter – (**sputter**), Quick talking due to excitement in a confused manner

Spoil – To ruin or damage

Spondylosis *(spinal osteoarthritis)* – This is a degenerative disorder; the general wear and tear that occurs in the human joints and bones of the spine as people get older; If and when this condition occurs in the zygapophysial joints, it can be considered facet syndrome. If severe, it may cause pressure on nerve roots with subsequent sensory and/or motor disturbances, such as pain, paresthesia, or muscle weakness in the limbs. (***See -Zygapophysial joints***).

Spouse –A wife or husband to someone

Spina Bifida – A congenital disorder caused by the incomplete closing of the embryonic neural tube

***Spinal Arthritis* (*Cervical Spondylosis*)** - A term used in the medical profession for the general wear and tear that takes place in the joints and bones of the spine as a person gets older; changes that comes with ageing affecting the neck (*cervical spine*)

Spinal cord - Is the Central Nervous System connected to the brain; from the brain, the cord runs down the back connecting to the peripheral nervous system

Spirometry - Is a common office test used to diagnose asthma, chronic obstructive pulmonary disease (COPD) and certain other conditions that affect breathing. It is used to check how well the lungs are performing once you're being treated for a chronic lung condition.

Spirometry measures how much air that someone can inhale and exhale. It measures how fast you can exhale. Spirometry values below average indicate the lungs aren't working as well as they should.

Spleen – Organ of the body close to the stomach; bad- tempered someone

Sputter – (***splutter***) Speaking in a confused manner due to excitement

Squamous-cell carcinoma or **Squamous cell cancer** (**SCC** or **SqCC**) - a cancer of a kind of epithelial cell, the Squamous cell

These cells are the main part of the epidermis of the skin, and this cancer is one of the major forms of skin: sometimes referred to as "**epidermoid carcinoma**" and "**Squamous cell epithelioma**"; symptoms are highly variable depending on the involved organs; SCC occurs as a form of cancer in diverse tissues, including the lips, mouth, oesophagus, urinary bladder, prostate, lung, vagina, and cervix, among others.

Squint – (strabismus Squint) A condition in which the eyes are not properly aligned; a misalignment of both eyes where both eyes are not looking in the same direction

Staff –Employed persons in a business or organisation

Stagger – Swaying from side to side

Standard precautions – Guide lines to prevent or reduce the spread of infection among people

Statement – Presentation in considered words – Implied correctness of a fact, position or problem

Statute – A kind of written law; an Act of Parliament

Stereotype – Generalised ideas being always misinformed about the ways that people from different backgrounds behave and feel; justified belief about individuals or group in society

Sterile – Free from bacterium (*pathogens*)

Sterilise –To make free from bacteria; to make incapable of reproduction

Stevens Johnson Syndrome(SJS): is a life-threatening skin condition, in which cell death causes the epidermis to separate from the dermis; a rare, serious disorder of your skin and mucous membranes. It's usually a reaction to a medication or an infection.

Stigma – The characteristics which causes a person to be disapproved of socially, avoided or rejected; a social disgrace

Stink – To emit a strong offensive odour

Stomach – An organ between the oesophagus and the small intestine. It has three tasks; It secretes gastric juice into the lumen, temporarily stores and mixes food

Strain –To cause one-self injury by over working

Stratum spinosum - The stratum spinosum is the fourth layer of human epidermis, which is the outermost portion of the skin.

Streptococcus – Group of bacteria that cause diseases **(*pneumonia*)**

Stress management – These are ways devised by individuals to overcome stress in their everyday lifestyle

Stress – Feelings of worry tension or pressure; stress is very difficult to define because it affects each person in different ways – signs of stress and its symptoms; ***Physical Stress*** (*aches and pains, dizziness, rapid heartbeat*): ***Emotional Stress*** (*short temper, depression, unable to relax*)***: Intellectual Stress*** (*Cognitive*) (*memory problems, constant worries, unable to concentrate*): ***Behavioural Stress*** (*use of drugs /alcohol, eating more or less, sleeping more or less, nervousness*)

Stroke –Sudden attack by paralysis; rapid loss of brain function due to disturbance in the blood flow to part of the brain. Sometimes called a "brain attack"

Stubborn –To carry on in their determined way; someone not willing to surrender

Subconsciously - Not wholly conscious; partially or imperfectly conscious; happening at a level without having full conscious thought or full awareness

Substance abuse – The miss use of drugs, solvents and alcohol in such a way that it causes problems to the health

Subcutaneous fluids - Subcutaneous fluid is a mixture of water, dextrose, and saline that is given to people via a needle under the skin to prevent; it is given to patients in order to prevent or treat dehydration

Suck –To pull with force into the mouth

Sudden –Unexpected happenings

Sugar – A sweet substance from the sugar beet root or sugar cane

Suicide –To take one's own life

Supernumerary – A staff member or trainee, not included in the care of service user or vulnerable tenants

Supervision – The practice of regular meetings with a senior member of staff, setting goals and professional development

Support – Supply with means to live

Surrogate-A substitute, a substitute mother or father

Suture – Stitching of wound; surgical stitch

Swab - A small piece of absorbent material attached to the end of a stick used for cleansing or applying medicine.

Swallow –Intake of food or water through the throat to the stomach

Swear – To curse

Sweat – Perspiration; moisture from the skin

Sweat glands –Excretion through the dermis of the skin to regulate body temperature

Swell –To bulge out; to become bigger

Swyer syndrome (*XY gonadal dysgenesis*) – This is a type of hypogonadism (*an interrupted "stage 1" in puberty;* deficiency of sex hormones that can result in defective primary, secondary or both sexual development and withdrawal effects in a person (e.g., premature menopause) in adults whose karyotype (*the appearance of the chromosomes in a somatic cell of an individual or species, with reference to their number, size, shape, etc*) is 46, XY. The person is externally female with streak gonads, and left untreated, will not experience puberty. Such gonads are typically surgically removed (as they have a significant risk of developing tumors) and a typical medical treatment would include hormone replacement therapy with female hormones.

Sympathetic Nervous System - The sympathetic nervous system (SNS) is part of the autonomic nervous system (ANS), it includes the parasympathetic nervous system (PNS). The sympathetic nervous system activates what is often termed the fight or flight response

Sympathy –To feel with the other person; an expression of pity for another

Symptoms – An indication of something; changes in appearance of ill health

Syndrome - group or recognizable pattern of symptoms; abnormalities that indicate a particular trait or disease.

Syphilis –Syphilis is a sexually transmitted infection caused by the spirochete bacterium Treponema pallidum subspecies pallidum; a sexually transmitted disease (STD) or sexually transmitted infection (STI) that, when left untreated, can progress to a late stage. It infects the genital area, lips, mouth, or anus of both men and women

Syringe – A tube used by doctors for injecting or withdrawing fluids; it consists of a glass, metal, or hard rubber tube with narrow outlet and fitted with either a piston or a rubber bulb; a device used to inject fluids into or withdraw them from something

Systemic Circulation - Refers to the part of the circulatory system in which the blood leaves the heart, services the body's cells, and then re-enters the heart. Blood leaves through the left ventricle to the aorta, the body's largest artery.

It supplies nourishment to all of the tissue located throughout the body, with the exception of the heart and lungs because they have their own systems.

Systemic circulation is a major part of the overall circulatory system

Systemic circulation

Synovial joints – Known as (diarthrosis), is the most common and most movable type of joint in the body of a mammal; where two bones meet. Ligaments and tendons hold joints together; there are 6 (six) types of synovial joints namely, gliding, hinge, pivot condyloid, saddle and ball and socket

Syrup – (Latin: *sirupus*) or (sirup (from Arabic: شراب‎; sharab) A thick, sweet, sticky liquid with sugar base, natural or artificial flavourings and water; any thick and sweet liquid prepared for table use from molasses, glucose, etc.

Systemic infection - Is that infection which affects a number of organs and tissues, or affects the body as a whole

"*T*rust in the Lord with all your heart and do not rely on your own understanding." (Proverbs 3:5)

Table – A flat surface standing supported on wooden or metal legs; columns laid out with figures or letters

Table cloth – Cloth used for covering the table

Tablet – A drug in solid form; a flat, small piece of stone with cut words

Tachycardia - A heartbeat that is too fast - above 100 beats per minute in adults

Tack – A short nail with flat head

Tactile – The sense of touch

Take –To grasp

Talk – To speak; discussion

Tamper –To interfere

Tank – Container used for containing fluids, can come in various sizes

Tantrum – Outburst of bad temper

Tape – Narrow strip of material used for binding, comes in various width, colours and sizes

Tape worm – Tapeworms are parasites that live in the intestines it is a ribbon-like flat worm;

It gets into the body via contaminated food; a tapeworm can easily survive and thrive indefinitely inside a human and can range in length from 1/250 of an inch (.0063 cm) to 50 feet (15.23 meters)

Tapioca –Starchy food extracted from cassava plant, commonly grown in the tropics

Target group / audience - A group of people that an advertising campaign is focused on selling to

Task – A specific duty assigned to a person

Taste – This the sensation produced when a substance in the mouth reacts to distinguish the flavour taken into the mouth

Taste buds – Sensitive cells on the tongue

Tattoo – Colourful patterns made on the skin by pricking it

Tea –Specially selected dry shrubs of leaves, made into drink by adding hot water

Team – People working together as in sports or hospitals, care homes, etc.

Teamwork – A group of persons working together to achieve same goals and objectives to effectiveness

Tease –To annoy or irritate

Technology Support – Using computer to produce voice and any other electronic type of sounds or communication aids

Teeth – Hard enamel protruding from the gum, used for chewing and biting

Teetotal – Someone who never drinks alcohol

Telephone –An instrument used for communicating by means of radio waves

Temperament –A quick mood swing

Temperature –The degree of hot or cold in measurement of degrees Celsius

Tempt –To entice; persuade

Tenants – People who pay rent for their lodging

Tender points - Specific places on the neck, shoulders, back, hips, arms, and legs. These points hurt when pressure is put on them

Tendon –A band that connects a muscle to a bone in the body

Tennis elbow – When tendons coming from the muscle of the forearm become inflamed at the point where they join the epicondyle **- *see epicondyle;*** irritations of the tendons of the forearm due to minor injuries or by over using the muscles.

Termagant – A violent and bad tempered woman

Terms - Conditions or stipulations limiting what is proposed to be done

Terminal Illness –A physical disorder; an advance stage of illness resulting in death

Termite – They infest house in colonies, similar to a bee hive or an ant's nest

Tertiary care –Care offered through special hospital services e.g. neurosurgery and cancer hospitals

Testicle - Testicles "Balls" suspended inside a skin sack underneath a male's penis

Testosterone - is a steroid hormone from the androgen group and is found in mammals, reptiles, birds and other vertebrates. It is secreted primarily in the testicles of males and the ovaries of females: small amounts are also secreted by the adrenal glands. It is the principal male sex hormone and an anabolic steroid

Testosterone therapy - is a treatment used for men whose testes do not produce enough testosterone (the main male sex hormone). It may be due to absence, injury, or disease.

Tetanus – (lockjaw), is an infectious disease caused by the bacteria Clostridium tetani

Tetraplegia – (quadriplegia), A Spinal Cord Injury (SCI), is paralysis caused by illness or injury to a human that results in the partial or total loss of use of all their limbs

Thalassaemia – Group of inherited blood disorders that affect the body's ability to create red blood cells; **Thalassaemia (*inherited blood disorder*)** - Originated in the Mediterranean region it has also been reported in hanging victims.

Thalidomide – is an anti-nausea and sedative drug that was introduced in the late 1950s to be used as a sleeping pill but it caused severe fatal malformations and deformities on babies born during that period

The Equality Act 2010 – This aims to protect and prevent previous legislation associated with race, gender, disability, age and discrimination in work, education, services, facilities, etc.; it states that all family members, service users and vulnerable tenants are fairly treated and are able to gain access and benefits from services and that health professionals and colleagues are treated equal

The Health Board - authorities in urban areas of England and Wales from 1848 to 1894. Formed in response to cholera epidemics and were given powers to control sewers, clean the streets, regulate slaughterhouses and ensure the proper supply of water to various districts. The boards were eventually merged with the corporations of municipal boroughs in 1873, which became urban districts in 1894.

Therapeutic - The treatment of disease or disorders by remedial agents or methods

Therapy – A process used to help people to overcome social difficulties e.g. counselling, physical exercise, etc.

Thermograph –Instrument used to measure the amount of heat produced in parts of the human body; used for detecting tumours

Thief – Someone who steals

Thinking – Bringing self-awareness to a maximum height and considerations from all points of view

Thoracic Aortic Aneurysms (*TAA*) - These are found within the chest; these are further classified as ascending, aortic arch, or descending aneurysms: Thoracoabdominal aortic aneurysms involve both the thoracic and abdominal aorta. There are other classifications that might help treatment - **Risk factors**-*Hypertension, Tobacco use, Peripheral vascular disease*

Throat – (Pharynx), the passage that connects the wind pipe, mouth, nose and oesophagus

Thrombosis – Blood clot; the formation of a clot in the blood vessel

Thymus –The gland near to the base of the neck

Thyroid glands – The gland in the neck where the larynx meets the trachea that produces hormones called thyroxin

TIA – (*See Transient Ischemic Attack*)

Tibia – The larger of the two bones between the knee and ankle

Tic disorder – This was once viewed as emotional factors. Increased risk for depression and other mood disorders, as well as anxiety disorders; Tics are disorders, the core symptom shared by brief tic disorder, vocal tic disorder, and Tourette's disorder. It is the severity and course that distinguishes these disorders from one another.

Tidal air – Air breathed in and out of the lungs during the cycle of breathing

Tissue – Group of independent cells in the body

Tone of voice - The quality of a person's voice; the tone and manner by which the person speaks

Tongue – The organ attached to the base of the mouth (speech, chewing and swallowing); the tongue is anchored to the mouth by webs of tough tissue and mucosa. Its tongue is a muscular organ in the mouth, covered with mucosa (a moist, pink tissue) with its rough textured tiny bumps called papillae. The tongue is vital for chewing, swallowing of food, as well as for speech. It is responsible for the four common tastes known as sweet, sour, bitter, and salty

Tonsil –These are two masses of tissues at the back end of the throat

Top coat –A lightweight overcoat worn as an outermost garment

Torment –An affliction; great bodily pain: to be tormented

Torture – Inflicting severe pain as punishment

Torch –A hand held light operated by battery

Touch –To come in contact; laying hands on...

Tourette's syndrome - is a compulsive **tic disorder** (*increased risk for depression and other mood disorders, as well as anxiety disorders*) and the most common tics are of eye blinking, coughing, throat clearing, sniffing, and facial movements; variant of such a disorder is that it is associated with compulsive masturbation, intrusive thoughts, there could be a family history of Tourette's syndrome or similar disorder.

Transfer –To move from one place to another

Transient Ischemic Attack – TIA: transient ischemic attack, is a "mini stroke" that occurs when a blood clot blocks an artery for a short time. The difference between a stroke and TIA is that the blockage with TIA is transient (temporary). The symptoms vary with different people; across the brain regions; TIA is caused by a clot; the difference between a stroke and TIA is that with TIA the blockage is temporary. TIA symptoms occur quickly and last for a relatively short time; TIAs can last for less than five minutes; the average duration is about a minute. Usually, it causes no permanent injury to the brain.

Transplant – To remove and plant or graft in another place

Tranquillisers – Drugs used to calm the nerves (*see - Sedative*)

Trauma –Shock; A serious injury as from violence; physical damage

Treatment –Behaviour towards a person; management in the application of medicines,

Trespass –An unlawful entry into a place

Trolley –A shelved stand on wheels used for serving tea and cakes

Trouble –Causing an uneasiness or inconveniences

Truculent – Someone very aggressive in behaviour; threatening

Tuberculosis – (*Mycobacterium tuberculosis*) lung infections; (TB) the deadliest bacterial disease in history; an infectious disease of the lungs

Tumour –A swelling; an abnormal growth of tissues in the body

Tutor - an instructor who gives private lessons; a tutor is not to be confused with a teacher who is employed in the *education* of groups

Two-ways – A set of directions

Types of Cancer - Bladder cancer; Melanoma; Brain cancer; Non-Hodgkin lymphoma; Breast cancer; Ovarian cancer; Cervical cancer; Pancreatic cancer; Colorectal cancer; Prostate cancer; Oesophageal cancer; Skin cancer; Kidney cancer; Thyroid cancer; Liver cancer; Uterine cancer; Lung cancer; Leukaemia.

Typhoid Fever - **(**Enteric fever**)**; disease is caused by a bacteria called Salmonella typhi. It affects all age groups; poor hygiene conditions, open sanitation habits, flies, sale of exposed food. The disease is transmitted from human to human via food or drinking water

"*U*ntil the dawn arrives and the shadows flee, turn, my beloved— be like a gazelle or a young stag on the mountain gorges." (Song of Solomon 2:17)

Ugly – Unpleasant

Ulcer – An open sore with pus; peptic ulcer disease (PUD), is the most common ulcer of an area of the gastrointestinal tract that is usually acidic and thus a burning stomach pain is the most common symptom

Ulcus molle – (*Chancroid*) - Is a bacterial sexually transmitted infection characterized by painful sores on the genitalia; it is known to spread from one individual to another solely through sexual contact.

Ulna – (elbow bone) is one of the two long bones in the forearm, the other being the radius

Ultramicroscopic - An instrument that uses special scattering features to detect the position of objects that are too small to be seen by a normal microscope

Ultrasound is an oscillating sound pressure wave with a frequency greater than the upper limit of the human hearing range; It is used in many different fields. The Ultrasound imaging known as ultrasound scanning or sonography involves the use of a small transducer (probe) and ultrasound gel to expose the body to high-frequency sound waves. Ultrasound is a useful way of examining many of the body's internal organs

Ultraviolet – *(UV)* A wavelength shorter than that of visible light, but longer than X-rays,

Umpire –Someone selected to sort out disputes about the application of settled rules; person put in charge to officiate a game

Unanimous – Having the agreement and consent of everyone

Unapproachable –Not easy to reach

Uncared- for – Neglected; not looked after

Uncle – The brother of someone's father or mother

Unconscious – The lack of response; not knowing

Undaunted – Fearless

Undergarments –These are clothes worn under other clothes

Undress – To take one's clothes off

Understanding - The process of comprehending or the knowledge of a specific thing or practice; a state of agreement

Uneasy – To be restless

Unisex – For the use of both male and female

Uniform – Clothing worn by a group of staff as a kind of official identity to recognition

United Nations Declaration of Human Rights (Article 8) - Everyone has the right to an effective remedy by the competent national tribunals for acts violating the fundamental rights granted him by the constitution or by law.

Unstable diabetes - A type of diabetes; when a person's blood sugar level often swings quickly from high to low and from low to high, it is known as "brittle diabetes"

Untimely – To occur unexpectedly

Unsafe practice – The type of care that is unsafe and put its service users at risks

Upset – As in distress

Uproar – A disturbance by making loud noise

Urinalysis -Is an array of tests performed on urine, and one of the most common methods of medical diagnosis.

Urinary tract infection (*UTI*) - Is an infection in any part of the urinary system, kidneys, urethras, bladder and urethra; the bladder and the urethra are most infected since they are of the lower urinary tract.

Urine –Acidic fluids passed out of animals and human beings

Urologists are doctors specially trained to treat problems of the male and female urinary systems and the male sex organs

Uterine Cancer - Uterine cancer usually occurs after menopause; there are different types of uterine cancer. The most common type starts in the endometrium, the lining of the uterus. This type of cancer is sometimes called endometrial cancer. The symptoms of uterine cancer include, Unusual vaginal bleeding or discharge; Trouble urinating; Pelvic pain; Pain during intercourse

Uterus – A woman's womb

Unhappy - Not happy or joyful

"Violent person entices his neighbour, and leads him down a path that is terrible." (Proverbs 16:29)

Vacancy – An unoccupied post

Vaccinate – Protection against diseases by inoculation

Vacuous – To be empty; silly

Vacuity – The emptiness of the mind

Vagabond – Someone with no settled abode

Vagina – The front passage between the female thighs, connecting to the womb;

(from Latin vāgīna, literally "sheath") is a fibro muscular tubular tract which is a sex organ and has two main functions; sexual intercourse and childbirth

Valsalva Manoeuvre - is performed by moderately forceful attempted exhalation against a closed airway, usually done by closing one's mouth, pinching one's nose shut while pressing out as if blowing up a balloon

Valuable – Of special value

Values – The standards acceptable of someone or of a group

Vehement – A very violent and argumentative person

Vein - A vein is an elastic blood vessel that transports blood from various regions of the body to the heart; the system of branching vessels or tubes carrying blood from various parts of the body to the heart.

Venereal Disease *(VD)* – Sexually transmitted and contagious disease

Venipuncture –The surgical puncture of a vein especially for the withdrawal of blood or for the infusion of liquid substances directly into a vein

Verbal – Communicating using words

Verify – Confirming the truth

Vermin – Pest; a worthless and hateful person

Vernacular – The native dialect or language of a country

Vertigo – (*Latin – verto-whirling*) Dizziness; unsteadiness; sudden drop in blood pressure or problems with the inner ear.

Vesicle - A small fluid-filled blister as tiny as the top of a pin; as on the skin

Vex – To annoy

Viagra - is a prescription-only medicine used to treat Erectile Dysfunction (ED)

Victim – someone seriously injured or killed intentionally or by accident

Viral illness - Viral diseases are extremely widespread infections caused by microorganism. The most common type of viral disease is the common cold, caused by a viral infection of the upper respiratory tract (nose and throat). Some common viral diseases include: Chickenpox, Flu (influenza), Herpes, Human immunodeficiency virus (HIV/AIDS), Human papillomavirus (HPV), Infectious mononucleosis, Mumps, measles and rubella, Shingles, Viral gastroenteritis (stomach flu), Viral hepatitis, Viral meningitis, Viral pneumonia

Viral diseases are very contagious and spread from one person to another; when a virus enters the body and begins to multiply.

Contaminated ways that viruses can spread from one person to another are: -

> ➢ *Breathing in air-borne droplets*
> ➢ *Eating food or drinking water*
> ➢ *Having sexual contact with someone readily infected with a sexually transmitted virus*
> ➢ *Indirect transmission from one person to another by a virus host, such as a mosquito, tick, or field mouse*
> ➢ *Touching surfaces or body fluids readily contaminated*

Virus – Bacteria that can grow on body cells; micro-organism infecting other cells in the body. Micro-organism that reproduce within a host's cell causing sickness

Visit - A short time call; a reason to socialise or for business

Visual – The sense of sight

Vitiligo - Is a skin condition in which there is a loss of (pigments) brown colour from areas of the skin, resulting in irregular white patches that feel like normal skin; it is more noticeable in darker-skinned people because of the white patches against dark skin. It may appear at any age; with increased rate of the condition in some families. It is a condition that affects about 1 out of every 100 people

Vivisection (*Latin- vivus, meaning "alive", sectio, meaning "cutting"*) - is defined as surgery conducted for experimental purposes on a living organism, normally, animals with a central nervous system, to view living internal structure, e.g. mice.

Violence – Uncontrollable behaviour

Victor – Someone who wins a battle; a winner

Vulnerable – Someone that is more susceptible to harm; being likely to be harmed

Vocal Cords – A membrane in the larynx which produce sounds

Vocabulary – (< Latin *vocābulum* a word, a name, equivalent to) the kind of words used by or known to a particular people or group of persons: a word considered only as a combination of certain sounds orletters, without regard to meaning

Vocational - connected with an occupation

Volunteer – Offering of services on own accord; a service offered willingly and without pay

Vomit – Contents of the stomach being thrown out forcefully.

"Wisdom is found in the words of the discerning person, but the one who lacks wisdom will be disciplined." (Proverbs 10:13)

Wavelength – A distance in the line of a wave from one point to similar point

Wage – Reward as payment for a service rendered

Waist – That part of the body between the ribs and the hips, usually of the narrowest part of the central part of the body

Waist coat – A sleeveless upper-body garment worn over a dress or shirt

Wait - To remain in expectation

Waive – To voluntarily give up one's rights

Wake – To be awaken from sleep also it could be a ceremony associated with death

Walk – This is the movement of the feet by alternately putting one foot before the other in a slower pace than running

Wall – A structure of masonry or wood raised to enclose or divide an area

Wallet – A small flat case, used to carry personal items such as cash, credit cards or other documents

Ward – A room with many beds in a hospital; to take care of.

Warden – One having care of something: guardian,

Warm – Comfortable and agreeable degree of heat; a moderate degree of heat

Warm hearted – Kindness and generosity

Warn – Advice of impending danger possible anything else

Warrant – Authorization of sanction, given by a superior; justification

Wash – To cleanse by using water or other liquid, usually with soap or detergent

Wasp – Is typically defined as any insect of the order Hymenoptera and suborder Apocrita that is neither a bee nor an ant

Waste pipe –A pipe installed to take used water away, as from a sink to a gutter/ river

Watch – A time piece worn around the wrist or pinned to an apron worn by a nurse in hospital

Water – contains one oxygen and two hydrogen atoms; essential for most plant and animal life to survive

Wear – Material or other object on a person as cover

Weave – The interlacing of threads of the weft and the warp on a loom

Weep – To shed tears

Weight – The heaviness of something

Wellbeing – Is being happy, prosperous and healthy; the condition of being contented, healthy, or successful; welfare

Welfare – Money or help given by the government to assist the poor

Wet nurse – is a woman employed to breastfeed another's baby

Wheat – A type of grass known for its grain; the grain is processed for food

Wheel chair – The device propelled by motors or by the seated person that is unable to walk

Wheeze – A whistling type of continuous sound produced; sound produced in the respiratory airways during breathing.

Whiplash – A term used to describe neck injury caused by sudden vigorous movement of the head forwards, backwards or sideways. Most injuries are temporary, not serious and are usually detected early.

Whisky – Alcoholic drink produced from grains

Whistle-blowing – Someone who reveals wrong doing within an organisation to the public through the media and/or to a higher authority

Whitlow – or hand herpes; a painful infection that typically affects the fingers or thumbs.

Whooping cough (*Pertussis*) — is a highly contagious bacterial disease caused by *Bordetella pertussis*. It is called the cough of 100 days.

Its symptoms are initially mild then develop into severe coughing fits, which produce the namesake high-pitched "whoop" sound in infected babies and children, when they inhale air after coughing. It lasts for approximately six weeks before subsiding.

It can be prevented by vaccination; antibiotics are recommended because they shorten the duration of infection. It is known that the disease affects about 48.5 million people worldwide and kills about 295,000 people yearly.

Wicked – Spiteful or wicked as of a person

Widow – A woman whose husband is dead

Will – Document containing what is to be done after a person dies

William Syndrome - (WS or WMS; also Williams–Beuren syndrome or WBS) is characterized by a distinctive, "elfin" facial appearance; It is a rare genetic disorder that affects a child's growth, physical appearance, and cognitive development

Winch - is a mechanical device that is used to pull in (wind up) or let out (wind out) or otherwise adjust the "tension" of a rope or wire rope ("wire cable "used *in Ambulance helicopters*).

In its simplest form it consists of a spool and attached hand crank. There are different types of winches, namely, *Snubbing winch (a vertical spool with a ratchet mechanism), Wake skate winch (for many water sports enthusiasts), Glider winch (used by gliders), Air Winch (an air-powered version of a winch commonly used for the lifting and suspension of materials; used by the oil and gas companies)*

Withdrawal Symptoms – a surge of adrenalinea produced by the brain causes physical or emotional disorders, including nervousness, headaches, and insomnia, that occur when an individual who, upon the discontinuation or decrease of the intake of medications or recreational drugs

Work - Physical or mental efforts of activities directed toward the accomplishment of something; a form of paid employment

Widower – A man whose wife is dead

Wife – A female to whom one is married to

Willis-Ekbom disease - (WED) - is a disorder of part of the nervous system that causes an urge to move the legs. It usually interferes with sleep and a sleep disorder. Commonly diagnosed in women, can begin at any age and in young children; people affected are of middle-age or older. Treatment of disorder can cause insomnia. Also known as *Restless Legs Syndrome - (RLS)*

Work – Effort made to achieve results; employment

Workforce planning – The planning laid down by employers to mark the size of their staff and their working roles; including the management structure

Wound – Type of injury which could be tearing of the skin (an open wound) or (a closed wound) which could be an injury to someone's feelings; often occur as a result of an accident or injury through physical aggression; any cut or injury caused by falling or accidental

Wrap – To cover by wrapping things round the object

Wreath – An assortment of flowers and leaves or various materials made to resemble a ring

Wrinkle – Furrow in the skin as a result of aging or frowning

Wrist – The joint between the forearm and the hand

Write – The form of scribble of letters, words, or symbols on a surface such as paper

Writing – Setting down on paper (documenting) experiences and observations as seen on a day- to- day basis, sharing and bringing fresh understanding

"Be Thou e**X**alted Oh Lord My God"

X- Chromosome - is one of the two sex-determining chromosomes (allosomes) in many animal species, including mammals (the other is the Y- chromosomes); the sex chromosome associated with female characteristics in mammals

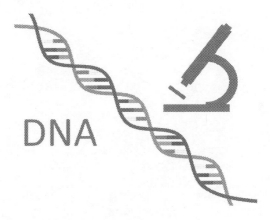

Xmas – Short form for Christmas

X-ray – Powerful invisible ray which penetrates the human flesh to photograph the bones

X-ray Examination - An X-ray examination is a painless procedure, used in creating images of the internal organs and bones to diagnose diseases. It is a small machine that outputs small amount of ionising radiation

Xenophobia – The fear of foreigners

***XY Gonadal Dysgenesis* (Swyer Syndrome)** – This is a type of hypogonadism (*an interrupted "stage 1" in puberty;* deficiency of sex hormones that can result in defective primary, secondary or both sexual development and withdrawal effects in a person (e.g., premature menopause) in adults whose karyotype (*the appearance of the chromosomes in a somatic cell of an individual or species, with reference to their number, size, shape, etc.*) is 46,

XY. The person is externally female with streak gonads, and left untreated, will not experience puberty. Such gonads are typically surgically removed (as they have a significant risk of developing tumors) and a typical medical treatment would include hormone replacement therapy with female hormones.

XYY Syndrome – Chromosome disorder (*see medical journal for more details*)

*"Y*ou, who are naive, discern wisdom! And you fools understand discernment! (Proverbs 8:5)

Y- Chromosome – (*DNA*) - One of the two sex chromosomes in humans (the other is the X chromosome); the sex chromosome associated with male characteristics in mammals

(1) The three major single chromosome mutations; deletion

(2) Duplication

(3) Inversion

Yankee – Sometimes shortened to *"Yank"*; outside the United States of America, someone from the USA is referred to as a Yankee

Yawn – To involuntarily open the mouth in a prolonged deep inhalation and sighing, usually from drowsiness or boredom

Yeah – Another word for Yes

Year – Time during which the Earth completes a single revolution around the sun; it consists of 365 days, 5 hours, 49 minutes, and 12 seconds; (in a leap year, 366 days) it is divided into 12 months and regarded in the Gregorian calendar as *beginning Jan. 1 and ending Dec. 31*

Yeast – (*as in infection*) Vaginal yeast infection symptoms like vaginal burning, itching, soreness, discharge, and pain during sex and urination; Yeast is a member of the fungus and a living organism in the air around us

Yeast – (*as in food production*) is a very helpful organism, especially in the making of wine, baking and brewing

Yell – To cry out loudly; scream, shout

Yellow – The colour of gold, butter, or ripe lemons. In the spectrum of visible light, and it is the traditional primary colour between green and orange in the visible spectrum

Yellow pages – The yellow big book in which business listings, phone numbers, addresses, maps, driving directions and more can be found

Yield – Reward; the return rate of an investment, income received

Yoga – Originated in ancient India with a view, an exercise to combat stress

Yoghourt – a fermented milk product (soy milk, nut milk, etc.) sometimes sweetened or flavoured

Yolk –The yellow part of an egg which feeds the developing embryo

Yorkshire pudding – also known as batter pudding; one of Britain's favourite and most popular dishes, usually made from batter and served with roast meat and gravy

Young Ladies – Female gender aged between 16 to 25 years of age.

Yuletide – (*Christmastide*), It is a religious festival observed by the Northern European peoples; Christmas season

Yuppie – A term referring to the upper middle class; a young educated adult, employed in a well-paying profession and who lives and works in a large city

"Zidele Amathambo" Give yourself up, bones as well
i.e. (take a chance) –
South African Ndebele Saying

Zeal – Energetic support

Zero Tolerance - means that certain actions will absolutely not be tolerated under any circumstances.

Generally used in reference to policies that spells out exactly which actions are forbidden. For example – National Health Service (NHS) - words, threats, or actions that are construed as bullying will be punished severely, Schools - many schools have a zero tolerance policy when it comes to bullying. A bully may even be suspended or expelled depending upon the intensity or the frequency of the behaviour.

Zero Tolerance Policy – (*in schools*) is a policy of punishing any infraction of a rule, regardless of accidental mistakes, ignorance, or extenuating circumstances. In schools or hospitals, common zero-tolerance policies concern possession or use of illicit drugs or weapons. Students, and sometimes staff, parents, and other visitors, who possess a banned item for any reason are always (if the policy is followed) to be punished.

Zero Tolerance Policy **(ZTP)** also apply to Health Services and other related public institutions.

Zika virus - a tropical disease which takes its name from the **Zika** forest in Uganda, transmitted by daytime-active Aedes mosquitoes. Commonly found in tropical regions of the world. It was isolated for the first time from humans in Nigeria; with evidence of human infection reported from other African countries such as the Central African Republic, Egypt, Gabon, Sierra Leone, Tanzania, and Uganda, as well as in parts of Asia including India,

V. K. Leigh

Indonesia, Malaysia, the Philippines, Thailand, and Vietnam such as Colombia, Ecuador, El Salvador, and Jamaica.

Zimmer – A walking support frame used in homes and hospitals; frame made of metal held in front as a walking aid

Zinc – A metal used to roof the top of buildings

Zygapophysial joints (*facet joints or z-joints*) - located on the back (posterior) of the spine on each side where two adjacent vertebrae meet. Facet joints provide stability and allow the spine to bend and twist. This joint contains cartilage between bones and surrounded by sac-like capsule that is filled with synovial fluid (*lubricating liquid that reduces the friction between bony surfaces with movement*). **Zygapophysial joint** (*facet joint*) - The joint between the superior articular process of one vertebra and the inferior articular process of the vertebra directly above it. There are two facet joints in each spinal motion segment.

Appreciation and thanks with reference to:-

Newworldencyclopedia.com
Wikipedia, the free encyclopaedias
https://en.wikipedia.org/wiki/File
Schizoaffective disorder
Wikimedia.org/Wikipedia/commons
www.emedicinehealth.com/script/
www.Amazon.cofee
https://Pixabay.com/en
the Wikimedia Commons. Information
Wikipedia, the free encyclopaedias
www.drugabuse.gov/sites/default/cocainepile
cocaine pile
http://en.wikipedia.org
https://pixabay.con/en
http://en.wikipedia.org/wiki/File:Diphalia_01.jpg

For diagrams and pictures use in the preparation of this book

Words of Wisdom

- "The Chameleon can never change its majestic steps, even if the bush is on fire"
- "The proverbial fly that refuses to listen to advice will accompany the coffin into the grave when it is buried"
- "A Gold fish has no hiding place"
- "There is no snake charmer that has not been bitten by a snake"
- "The evil that men do live after them"
- "Every day for the thief to have their ways but one day is for the owner of the property to catch him"
- "When God says yes, no man or creature can say no"
- "It is only a fool that will see blindness and open his eyes"
- "The ways of the wicked is like gloomy darkness; they do not know what causes them to stumble" (Proverbs 4:19)
- "Success is a matter of hanging on when others have given up"
- "You never lose until you quit trying"
- "The wicked runs when no one is in pursuit"
- "To be courageous with love; is one of the greatest gifts of life"
- "It is in the shelter of each other that people live"
- "Where there is great love, there are always miracles"
- "Prosperity is like a tender mother but blind, who spoils her children" (English Proverb)
- "The way is like a mountain path not soft grass but it goes upwards, forward, towards the sun"
- "For with great wisdom comes great frustration"
- "Whoever increases his knowledge merely increases his heartache."
- "Human Government Demonstrates Limitations of Wisdom"
- "Oppressors like unhappy people, because sadness exists within them"
- "To bear a child does not mean that you will lead the child on the right path of life"

- ➢ "Outside of Christ is crisis"
- ➢ "You have to expect things of yourself before you can do them." *By Michael Jordan, Basketball star*
- ➢ "A feast is made for laughter, and wine makes merry: but money answered all things".
- ➢ "There is flattery in friendship" – William Shakespeare
- ➢ "Beware of "**frienemies**" (enemies who pretend to be friends)"
- ➢ "He who collects ant infested wood, invites the lizards for a feast"
- ➢ "He that is down, need fear no fall"
- ➢ "For God gives those who please him wisdom, knowledge, and joy; but if a sinner becomes wealthy, God takes the wealth away from him and gives it to those who please him". (Ecclesiastes 2:26)
- ➢ "When the breeze blows, it exposes the bottom of the fowl"
- ➢ "When the crocodile leaves the river, the frogs go in to swim" – African saying
- ➢ "Understanding is deeper than knowledge; many people know but few people understand"
- ➢ "Words spoken cannot be taken back; they can be forgiven but not forgotten"
- ➢ "There's no amount of makeup that can cover up an ugly personality"
- ➢ "A tongue has no bones but it is the most powerful weapon in the human body; strong enough to break a heart. Be careful of what you say"
- ➢ "Honour and Hunger have never married successfully"
- ➢ "The devil you know is better than the angel you do not know"
- ➢ "Integrity is a virtue that man cannot buy"
- ➢ "The gift of a man will open the door to a man's greatness"
- ➢ "The gift of a man makes great way for him"
- ➢ "A house divided cannot stand"
- ➢ "If a herbalist should know all the names of the herbs in the evil forest, death would pay homage to him"

> "The strongest action for any person is to love them self, be whom they are and shine among those who never believed that they could make it."

> "When people walk away from you..., let them go... Your destiny is never tied to anyone who leaves you; it does not mean that they are bad people, it just means that their part in your life is over"

> "Change is never easy, you fight to retain and fight to let go"

> "If people say something **BAD** about you, judge you as if they know you, do not get easily offended but remember that "***Dogs Bark If They Do Not Know the Person***"

Based on Facts

➢ Human beings are born with 300 bones in their body but because they fused together, an adult have 206 bones
➢ Flamingos get their distinctive colours from the kind of food that they eat
➢ People do not sneeze while they are fully asleep
➢ The human brain is only two percent (2%) of a person's body weight but requires about twenty percent (20%) of its oxygen and calories
➢ Women blink twice as often as men
➢ Only twenty percent (20%) of Americans have travelling passports
➢ Almost half of the population of Manhattan in America, live alone
➢ The human heart beats more than 100.000 times in one day
➢ Black cats are considered bad luck in the United States of America and in Parts of Africa (Nigeria) but are considered Good Luck in Japan
➢ Turtles breathe through their rear ends
➢ Dolphins sleep with one eye open
➢ There are more chickens than people in the world
➢ In China, more pigs are produced than any other countries of the world
➢ Bullfrogs do not sleep
➢ Butterflies taste with their feet
➢ Seventy percent (70%) of the red meat eaten in the world is goat meat
➢ The hair on a polar bear is NOT white but transparent
➢ Ninety-seven percent (97%) of the earth's water is undrinkable
➢ Pineapple is not a single fruit but berries that are fused together
➢ The Eiffel tower in France gets about six inches (6 ins.), fifteen centimetres (15 cm) taller in the summer's heat

- ➢ The average depth of the ocean is 2.7 miles (4.3 km)
- ➢ Lightning strikes the earth about eight (8) million times in one day
- ➢ "SILENT" and "LISTEN" use exactly the same letters
- ➢ "Money can never buy contentment"
- ➢ "Quality goes in before the name"
- ➢ "Wedding Ring, a perfect circle without a beginning or an end; the symbol of the love that will last forever"
- ➢ Pronouncement on a newly married woman – "May your life be filled with love, light and laughter"
- ➢ "Time is like a river. The same water cannot be touched twice because the flow that has passed will never pass again. A clock cannot be turned back"
- ➢ "Never go to sleep angry because you never know if you or the person that you are angry at will wake up the next morning"
- ➢ "Thoughts can become words; words become actions; actions become habits; habits become character and character becomes destiny"
- ➢ "Be careful how you treat people for you never know where you will one-day meet them"
- ➢ "The brain is 80% formed of water"

Now you know!!!

Information *for* Sheltered Housing / Extra Care Tenants

This is additional useful information for vulnerable clients especially those who live in care homes and sheltered housing.

Make a written legal will to protect your assets and family members

General Useful Information

This information is to help older people make decisions about themselves, the type of care and support that they may need.

More details can be obtained from the partnership website between the Elderly Accommodation Counsel (EAC) and other agencies

(Detailed information can be obtained from housing organisations)

Death of a Tenant

When someone dies in sheltered accommodation, **what happens?**

Especially if they've had no-one living with them.

The tenancy will not end immediately after a tenant dies unless there is a written will from the late tenant; but if there is no written will, the only people who can end it are: -

An Executor –

This is someone named in a tenant's will as the person who will deal with the deceased possessions.

An Administrator –

This is someone who has applied to the Probate Registry and / or has the "Grant Probate" – letter of administration.

A Next of Kin, who is <u>NOT</u> one of the above, cannot end the tenancy after the death of a tenant.

The (*Organisation's*) Legal Team

Where there is no Executor or Administrator, by law in any Council based sheltered accommodation, a Notice to Quit on the Public Trustee is served. This means that the tenancy will end between the periods of four to six weeks after the notice is served.

During this time, if letters of Administration are obtained before the four /six weeks are up, then the Administrator can end the tenancy.

A closing statement of the rent account is then prepared and sent to the person who will provide the deceased details for this purpose.

Housing Benefit

When a person dies while receiving benefits, all the benefits will automatically stop on the day and date of their death.

This is because the statutory regulations state that payments of benefits should stop immediately on the date of their death and therefore will not cover the charges between the date of the death and the tenancy ending.

ALL benefits that were entitled to the deceased person(s) will stop immediately on the day of their death and the charges will be made against the estate once the tenancy is ended.

Access into the Property

If any individual already has keys, the housing authority will not get involve in who can get into the property.

If nobody has keys to the property, the housing authority can only give keys to the person who is the executor / administrator.

It is advised that the property is cleared within four weeks and all keys returned to the housing authority.

If there are unwanted goods left behind by the deceased family, a form is completed requesting for the disposal of the unwanted items free of charge by the housing authority.

If someone has applied for letters of administration and has not received them by the time the tenancy ends, the housing authority will store any goods for a reasonable amount of time; until these have been issued and the goods claimed.

Tenants Information

It is very important that all details relating to tenants are up-to-date and current (Support Plans and Risk assessments); prepared by the sheltered scheme manager and the tenant while still alive.

These details must be updated as and when new information is received from relatives or any other relevant source that is vital to go on the tenants file.

If the tenant is new and moving in, especially if it happens to be on a weekend, it is best practice to prepare some emergency information for Telecare – Emergency Team - to work from until the manager returns to work on the following Monday and if the staff will be away, it is advisable to leave a written note for the covering sheltered scheme manager and inform the up line Manger – they will know what to do.

As a sheltered scheme manager or sheltered scheme coordinator; do not **FAX** the tenants' information because of its high sensitive nature; send it by

E-mail – intranet - to Telecare; this way, there is an assurance and evidence of a record that can be easily referred to and as a responsible Sheltered Scheme Manager, you are covered.

All tenants' information MUST be securely preserved and kept confidential.

Decanting (This comes around every now and then)

What is Decanting?

Decanting is a process that takes place when works are required or being carried out to your Council home and it is decided that it is not safe for you to stay whilst they are carried out. You may be asked you to move to an alternative accommodation. This is called "decanting".

Anyone could find themselves being decanted in an emergency, if their homes are not fit to live in e.g. flood, fire, etc.

Sheltered Scheme Managers could find themselves involved in decanting process.

This happens when one or more schemes are being closed down and the tenants are being transferred from one sheltered scheme to another.

Quality Assessment Framework (QAF) Objectives

This is also known as the Community Care Legislation which applies to the tenants.

It emphasizes on Care within the community and encourages supported living in the community.

It promotes good services through good assessment of needs, taking into consideration ethnicity and cultural issues in the delivery service.

Through this act, older people are expected to play more active roles as well as exercising choice and control which they have been given through Supporting People's code of practice.

The core objectives are: Support Plans, Risk Assessment, Equality and Diversity, Security, Health and Safety, Confidentiality, Safeguarding, Complaints, Protection from Abuse, Keeping of Professional Boundaries and Advocacy.

Fair access, Tenants Involvement and Empowerment

Equality and diversity in the workplace

Treating people equally is essential to being an effective and productive organisation. The introduction of equalities legislation has resulted in the removal of many inequalities within the workplace. Councils now aim to make equalities central to their employment practice. This is to ensure a diverse workforce that reflects the communities they serve.

The EU's Council Directive (2000/78/EC) establishes a general framework for equal treatment in employment and occupation. It covers discrimination against people at work on the grounds of age, disability, sexual orientation and religion or belief. The Local Government Employers (LGE) works with councils' associations, regional employers and other bodies. The LGE leads and solves problems on pay, pensions, the employment contract, and offers relevant guidance.

About the Author

V.K. Leigh is a professional business manager and consultant in management training with vast experience and knowledge in the field of social care

V.K. Leigh is an experienced "hands on deck", "sheltered scheme manager"; he has had an inside, practical knowledge, experience and the privileges of working with older persons in the United Kingdom; as a worker and consultant in some of the largest, well known housing organisations and local authorities in the country; both within the private and public sectors, through the employment agencies and as a contracted worker.

Combining the years of practical experience and knowledge in sheltered housing, the Care and Mental health sectors, V.K. Leigh will undoubtedly be ranked as a professional in these sectors.

In writing this *"A 2 Z book in health and social care"* the words and contents go back to the day-to-day running of sheltered housing sectors of Supported Housing for the Older Persons, Social Care, Hospitals, Community Care, Day Care Centres and related sectors; sharing knowledge with anyone and everyone who reads or uses this book.

This book is useable as information based; instantly giving the meanings of unexpected words of doubts or puzzles to the human brain while working; it could also be used as a working manual or referencing "hand-held" book for quick answers to information and answers needed at that particular time.

V.K. Leigh continues to make research into ways of "how to improve the lifestyles of the vulnerable persons of all ages, especially the older persons"; not only in the United Kingdom but all over the world; searching for new ways and ideas of how to improve the vulnerable lifestyles for everyone.

V.K. Leigh is a professional. Computer Systems Analyst, Credit and Financial Analyst, published author and member to some of the world's known management organisations.

Other publications by this author

- **General Guide to Management in Sheltered Housing**

- **General Extra Care – *The Full Facts* -**

- ***Guide to Scheme Managers' Operations***

- ***Why Me?***

Obtainable from

www.xlibrispublishing.co.uk

www.amazon.com

www.barnesandnoble.com

And

Other good bookshops